Praise for *Lean Supply Chain and Logistics Management,* by Paul Myerson

"The Author has provided us with a comprehensive and thorough examination of Lean principles applied to improving the efficiency and effectiveness of supply chains and logistics. Managers who may have felt overwhelmed with the challenges and complexities of logistics and supply chain management in today's global environment will benefit from Paul Myerson's discussion of Lean. He has presented the material in an easy-to-understand style, and he addresses the basics as well as the advanced dimensions of Lean Management. He provides perspectives and insights which will add to the understanding of the power of Lean not only for those with limited exposure but also for those with experience in the area. His own experience and understanding is reflected in his discussion of Lean tools and the methodology for implementation of the Lean approach. The documented benchmarks for success and the many examples help explicate the complexities for the reader.

The book is organized and written so that it will be useful as an introduction to the field and also as a reference when special challenges arise for the practicing manager. The chapter on metrics and measurements should be particularly useful, as well as the chapter discussing global supply chains. The book was written primarily for managers in the field, but it could be a valuable resource for a collegiate level course or a professional workshop. Paul Myerson's book is a valuable contribution to Logistics and Supply Chain Management at an important time for companies facing the competitive challenges of the current global economy."

DR. JOHN J. COYLE
Professor Emeritus of Logistics and Supply Chain Management
Department of Supply Chain and Information Systems Smeal College of Business
Pennsylvania State University
University Park, Pennsylvania

"Supply chain and logistics management are more and more a driving force in the marketplace. Paul Myerson's 30+ years of experience as a manager, consultant, and teacher are very effectively put to use in his book that shows readers how to 'make things happen' instead of 'wondering what happened.'

Myerson provides readers with an orderly framework for achieving, maintaining, and improving a supply chain's competitiveness. With the consumer in mind, the emphasis is on continually eliminating non-value adding activities in a supply chain so that the buyer gets true value for the price paid.

To realize the benefits of a lean supply chain and logistics management takes the right frame of mind and organizational culture combined with the use of appropriate concepts, tools, and use of evolving information technologies as presented in Myerson's book."

WILLIAM DEMPSEY

"The book is a must read for all supply chain managers seeking to drive down costs and improve profits. Get copies for your controller and all senior managers *before* any investment is made in your supply chain. This book lays it all out. Whether new or senior to the management of a supply chain, I strongly recommend this book."

RICHARD LANCIONI
Chair, Marketing & Supply Chain Management
Fox School of Business
Temple University
Philadelphia, Pennsylvania

About the Author

Paul Myerson has been a successful change catalyst for a variety of clients and organizations of all sizes. He has more than 25 years of experience in supply chain strategies, systems, and operations that have resulted in bottom-line improvements for companies such as General Electric, Unilever, and Church and Dwight (Arm & Hammer). Myerson holds an MBA in Physical Distribution from Temple University and a BS in Logistics from The Pennsylvania State University. He is currently Managing Partner at Logistics Planning Associates, LLC, a supply chain planning software and consulting business (www.psiplanner.com). Myerson serves as an adjunct professor at several universities, including Kean University and New Jersey City University. He is the author of a Windows-based Supply Chain Planning software, and co-author of a new Lean supply chain and logistics management simulation training game by ENNA (www.enna .com/lean_supplychain/).

Lean Supply Chain and Logistics Management

Paul Myerson

New York   Chicago   San Francisco
Lisbon   London   Madrid   Mexico City
Milan   New Delhi   San Juan
Seoul   Singapore   Sydney   Toronto

The **McGraw·Hill** Companies

Library of Congress Cataloging-in-Publication Data
Myerson, Paul.
  Lean supply chain and logistics management / Paul Myerson.
    p. cm.
  Includes index.
  ISBN 978-1-265-62966-3 (pbk.)
  1. Production management. 2. Manufacturing processes. 3. Production control.
  4. Lean manufacturing. I. Title.
  TS155.M847 2012
  658.5—dc23

                                                              2011044874

McGraw-Hill books are available at special quantity discounts to use as premiums and sales promotions, or for use in corporate training programs. To contact a representative please e-mail us at bulksales@mcgraw-hill.com.

Lean Supply Chain and Logistics Management

Copyright ©2012 by The McGraw-Hill Companies, Inc. All rights reserved. Except as permitted under the United States Copyright Act of 1976, no part of this publication may be reproduced or distributed in any form or by any means, or stored in a data base or retrieval system, without the prior written permission of the publisher.

1 2 3 4 5 6 7 8 9 0 DOC/DOC 1 9 8 7 6 5 4 3 2

ISBN 978-1-265-62966-3
MHID 1-26-562966-8

Sponsoring Editor Judy Bass	**Project Manager** Neha Rathor, Neuetype	**Indexer** Robert Swanson
Production Supervisor Pamela A. Pelton	**Copy Editor** Erica Orloff	**Art Director, Cover** Jeff Weeks
Editorial Supervisor David E. Fogarty	**Proofreader** Manisha Sinha	**Composition** Neuetype
Acquisition Coordinator Bridget Thoreson		

Information contained in this work has been obtained by The McGraw-Hill Companies, Inc. ("McGraw-Hill") from sources believed to be reliable. However, neither McGraw-Hill nor its authors guarantee the accuracy or completeness of any information published herein, and neither McGraw-Hill nor its authors shall be responsible for any errors, omissions, or damages arising out of use of this information. This work is published with the understanding that McGraw-Hill and its authors are supplying information but are not attempting to render engineering or other professional services. If such services are required, the assistance of an appropriate professional should be sought.

This book is dedicated to my brilliant and beautiful wife, Lynne, and wonderful son, Andrew. My wife always comforts and consoles, never complains or interferes, asks nothing, and endures all. She also writes my dedications.

CONTENTS

FOREWORD

The world of costs and cost management has always been an Achilles heel for logistics managers. They pay a lot of attention to costs, as do other managers in a firm—especially the accountants and financial managers who put pressure on the logistics manager to cut, cut, and cut. Ironically, while the pressure is intense, the accuracy of the cost information is often low. For many years, beginning in 1970, there was the concept of logistics cost leverage that held that a "one penny reduction in logistics costs would result in a four to five times" revenue impact on the bottom line. The result of this idea was that most firms thought they were cutting costs, but in reality, their efforts were only partially successful. In *Lean Supply Chain and Logistics Management,* Paul Myerson provides a valuable approach that all logistics managers can use to identify important cost areas in the supply chain, and offers techniques for how to reduce these costs over time.

In the first chapter, Myerson defines a "Lean supply chain" and explains why there is such an interest in Lean. In the second chapter, he discusses the "need for speed" and why reducing costs can impact a supply chain in a dramatic way. He points out that the Six Sigma approach can be very useful in transforming a modern supply chain into a Lean supply chain. In Chapter 3, he poses the question, "What is waste?" He goes on to discuss the different kinds of waste to look for in a supply chain. These include inventory, transportation, waiting, overproduction, over-processing, defect, and behavioral waste.

In Chapter 4, Myerson points out that you can't plan, source, deliver or return materials in a supply chain unless you have an accurate understanding of the system's cost structure. He develops this concept further, emphasizing that "you can't build a house without a solid foundation" and understanding of the work plan, the job layout and the workplace organization and standardization. Organizing for a Lean supply chain requires that a firm establish batch-size reduction, quick changeover, quality at the source, and work cells. In addition, Lean analytical tools are necessary for gathering, organizing and identifying problems in the supply chain.

Myerson discusses the need to apply some of the important techniques for cost reduction in Chapter 7, including "just-in-time inventories" (JIT), and illustrates how beneficial they are for achieving leanness. He provides examples of some of the leading Lean supply chains, including those of Wal-Mart and Dell. He focuses on the visibility and reliability in their supply chains and how "cross-docking" is employed to achieve lean ideals. Chapter 8 details how "Lean Thinking" can be used in operating a warehouse—from the assembling of orders to value stream mapping. Chapter 9 discusses global supply chains and the use of "Value Stream Mapping" to identify waste. Myerson points out the areas of potential waste in global supply chains, and where waste can be reduced. Of course, leanness cannot be achieved without the involvement of the whole organization, through Lean training, management support and the establishment of a Lean operational and organizational structure.

Teamwork is essential, and integrating the sales force and other operational groups like marketing and production are, as Myerson points out in Chapter 10, important keys for success. Chapter 11 details putting together a Lean plan to get started, comprising Value Stream Mapping and Kaizen events. The use of technology in achieving leanness is discussed in Chapter 12, including ERP, DRP, Advanced Planning and Scheduling systems, Warehouse Management Systems (WMS), the use of RFID, and Transportation Management Systems.

Chapter 13 emphasizes working together with a firm's suppliers and customers. The collaboration is clearly achieved through the use of EDI, e-commerce, efficient consumer response, quick response systems, collaborative planning, forecasting, and replenishment. Myerson suggests vendor-managed inventories as a way to achieve integration and leanness in a distribution channel between suppliers and customers. Knowing where Lean is taking place in a supply chain and determining whether the Lean goals have been achieved is discussed in Chapter 14, in his treatment of Metrics and Measurement. Myerson points out the importance of the key logistics metrics, including delivery reliability, responsiveness, flexibility, cost, and asset management. The concept of the balanced scorecard is presented, and the need for dashboards to display and control metrics is emphasized.

The last two chapters (15 and 16) focus on the need for proper training of employees and managers to achieve a Lean supply chain. Myerson discusses the important methods, tools, and tips to use in training and measuring the success of the training. He points out that identifying the barriers to supply chain integration should be considered, including human resources, organizational structure and relationships, technology, and alignment. The author concludes by discussing the future trends in Lean supply chain systems and the potential obstacles to Lean thinking in the supply chain.

The book also includes two Appendices that discuss examples of real-world Lean supply chains and extensive references for Lean opportunity assessment.

RICHARD LANCIONI
Chair, Marketing & Supply Chain Management
Fox School of Business, Temple University
Philadelphia, Pennsylvania

PREFACE

Over the past 30 years, we have seen a dramatic growth in the size and complexity of business supply chains, making them much more challenging and difficult to manage.

At the same time, technology has rapidly improved and is readily available to companies of all sizes to help enable these processes. This, of course, assumes the various supply chain processes are functioning optimally, which is often not the case.

Part of the reason for many of these suboptimal supply chains is that while we have seen an increased use of Lean tools in manufacturing during this same period, it has only been in the past 5 years or so that both Lean and Six Sigma have made their way into the supply chain and logistics function to help simplify and improve these processes.

As a result of the relatively short time Lean has been applied in this area, there has not been a lot written on this subject to help supply chain professionals find their way. Most of the books that have been written on the subject tend to be very narrow in scope either by function (e.g., focused on procurement only) or topic (e.g., primarily covering one Lean tool such as 5S-workplace organization or value stream mapping).

The purpose of this book is to give practitioners the necessary tools, methodology, and real-world examples to successfully implement Lean in their supply chain and logistics function.

This book is organized so that the reader can first gain some perspective on how the supply chain and Lean have evolved in recent years. This is followed by explanations and examples of both basic and advanced Lean tools, as well as specific implementation opportunities in the supply chain and logistics function. A Lean implementation methodology with critical success factors is then identified and described.

In addition to real-world examples throughout the book, there are also a number of case studies on the subject to gain insight into how others have successfully implemented Lean in their supply chain.

Additionally, this book comes with a Lean Opportunity Assessment tool (Appendix B) and training slides (mhprofessional.com/myerson) to help make the transition from understanding to implementing Lean in your company or your client's company.

PAUL A. MYERSON

CHAPTER 1

Introduction: Using Lean to Energize Your Supply Chain

The major trend of outsourcing manufacturing and procurement of materials began around 1980 when the term "supply chain" was coined, and with it came added attention to the importance of the supply chain and the logistics management field. This was a result of the added complexity, longer lead times, and increased risk involved in such a major paradigm shift.

Starting around the same time and with ever-increasing speed, major advances in technology occurred, such as *electronic data interchange* (EDI), *enterprise resource planning* (ERP), *supply chain management* (SCM), and *supply chain planning* (SCP) systems, along with the creation of the Internet and e-commerce. These advances helped to improve planning and management efficiency. These technologies have also led to what many are calling *mass customization*, in which product development, life cycle, and delivery times have become compressed. On top of this, the supply chain is at risk of disruption caused by political, financial, environmental, and other unplanned events. These types of disruptions seem to occur at an ever increasing rate.

Managing the supply chain with all of this change and disruption is challenging; however, there are tremendous opportunities as well. That is what this book is all about. We will discuss in detail what Lean is and how it can be a tool to improve your organization's supply chain and logistics performance.

What Is Lean?

First of all, let us examine what Lean is not. I do not know how many times I have visited a company to discuss Lean and the owner or president said something like, "We're already lean . . . we laid off 20 percent of the workforce."

1

Contrary to what some people think, Lean is not some kind of crash diet where a company sheds its fat quickly. While the net result is to do more with the same or less, that should not be the goal. If it is, then your Lean journey is destined to fail.

Lean, in a nutshell, is a team-based form of continuous improvement that focuses on identifying and eliminating "waste." Waste, in this case, is non-value-added activity from the viewpoint of the customer. Instead of a diet, Lean should be thought of as a long-term health program for your business. Consider it a way to add energy and vitality to your company in an increasingly competitive, unstable, and generally challenging environment.

Lean Failure

Over the many years that I have been helping companies become Lean, I have read (and heard) that well over 50 percent of Lean initiatives (at least in the United States) have failed. From my experience, I tend to think this is accurate. I believe that the primary reason for this is the lack of the proper culture to support the major—and in some cases, radical—changes required.

Lean initiatives fail for a variety of reasons. In many cases, management is not willing to give up some control to workers, dedicate resource time, or spend money for training and improvements. In other cases, Lean is just looked at as a "fad" that will go away or a short-term program.

Most U.S. companies seem hesitant to spend time and money on the training required to become leaner, and they lag far behind other countries in the amount of training, in general, provided for employees. While there are a few bright spots, such as General Electric, with its Croton-on-Hudson John F. Welch Leadership Development Center facility, most offer very little in the way of training and support. In general, the average annual training hours per employee in the United States is much less than most other developed countries.

Implementing Lean

Lean also requires both a top-down management commitment to change and a bottom-up groundswell of participation and ideas. Otherwise, it is a losing battle. The culture has to encourage and create a team-based continuous improvement mentality.

In the United States, the more common way to initiate change has been to have consultants come in who "borrow your watch to tell you what time it is" and then leave reams of reports, slide shows, etc. for the client to review and interpret on their own. A myriad of problems can result from this methodology, ranging from no direction for how to implement the improvements to a general lack of enthusiasm and support from the people being asked to change based on suggestions from an outsider.

A more effective way to implement Lean, I believe, is through a *train-do* method. In this way, the trainer or consultant is more like a facilitator who teaches the employees the basic concepts and tools, but lets them create and direct the activities (with a little management oversight, of course). That way, after the trainer or consultant is gone (which is always the case eventually), workers can continue on the Lean journey.

Middle and upper management still does much of the higher-level "out of the box" thinking, using tools described in this book, such as *value stream mapping* (VSM), which is used to analyze and design the flow of materials and information in order to help identify opportunities; however, input and actual implementation has to include everyone in the organization.

In many cases, there also seems to be a gap between general Lean concepts training and how to actually begin the improvement process (and to convince management that it can benefit the company's bottom line, not just help employee participation and morale). That is why doing a *Lean opportunity assessment* (LOA), discussed in Chap. 11 (and template included in Appendix B), is a great place to start.

Historically, Lean was applied to the manufacturing industry first (initially assembly line manufacturers, then other types), hence the term Lean Manufacturing, which is still the most common term used. Around 10 years ago, Lean began to be applied to the office for administrative processes (also known as Lean Office) and more recently to the supply chain and logistics function. As a result, today, many tend to refer to it as Lean Enterprise.

Supply Chain and Logistics Management Defined

As this book is focused on Lean in supply chain and logistics management, it is necessary to first define the scope of this function. There are many definitions and understandings of supply chain. Technically, supply chain management includes the logistics function, which includes the transportation and distribution areas. However, as some take a very narrow definition that is primarily focused on procurement and purchasing, I think it is important to define what I mean by it and what this book covers.

In fact, according to a 2011 article in *Inbound Logistics* entitled "Continuing Education—Making the Right Selection" by Perry A. Trunick, "Some of the most passionate debates in academic circles still center on what constitutes supply chain management and its place in the academic structure. Not surprisingly, that same debate rages in the commercial world." The article goes on to say that some people use the terms "logistics" and "supply chain" interchangeably; others feel that it is important for logistics to still have its own place [Trunick, 2011].

It is more effective, especially from a Lean perspective, to take a very *broad* view similar to what is defined by the Council of Supply Chain Management Professionals (CSCMP): "Supply chain management encompasses the planning and management of all activities involved in sourcing, procurement, conversion, and logistics management. It also includes the crucial components of coordination and collaboration with channel partners, which can be suppliers, intermediaries, third-party service providers, and customers."[www.cscmp.org, 2011]

Another way to look at it is the Supply Chain Operations Reference (SCOR) Model (Fig. 1.1) from the Supply Chain Council [www.supply-chain.org, 2011], which divides the supply chain into five management processes:

1. Plan
2. Source
3. Make
4. Deliver
5. Return

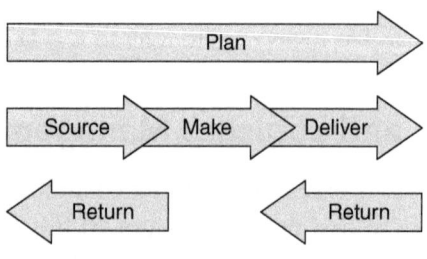

Source: Supply Chain Council (SCC).

Figure 1.1 SCOR Model.

All five management processes include management of risk, assets, inventory, metrics, performance against those metrics, business rules, and regulatory requirements. In addition, the processes include the functional responsibilities described below:

1. **Plan**—Balancing supply and demand [discussed in detail when describing the *sales and operations planning* (S&OP) process in Chap. 10]. These plans are communicated throughout the supply chain.
2. **Source**—The procurement of goods to meet demand. This includes identifying, selecting, and measuring performance of sources of supply, as well as delivering and receiving materials.
3. **Make**—The transformation process, taking raw materials and converting them into finished products.
4. **Deliver**—Resources to move materials along the supply chain, from suppliers to manufacturing and then to customers. Includes order management, warehousing, and shipping.
5. **Return**—The reverse logistics process for product or material that is returned, including repair, maintenance, and overhaul.

By taking this broader view of supply chain management, I include a variety of functional areas in my definition (and thus, potential Lean opportunities):

▲ **Information management (including generating and sharing customer, forecasting, and production information)**—This includes all of the information required to ensure that supply matches demand throughout the supply chain.

▲ **Procurement**—The acquisition of goods or services, which may include incremental functions beyond the purchase, such as expediting, transportation, quality, etc.

▲ **Inventory flow scheduling and control**—Responsibility for the accuracy, timeliness, and management of *maintenance repairs and operations* (MRO), raw material, *work-In-process* (WIP) and finished goods inventory.

▲ **Transportation systems operation and infrastructure**—The efficient movement of material and goods throughout the supply chain.

▲ **Distribution facilities management**—Storage of goods until needed by the customer (or manufacturing in the case of raw materials and components).

▲ **Customer service (including order management and fulfillment)**—Servicing the customer throughout the order process.

Why All the Interest in Lean Supply Chain Management?

Why is there so much attention paid to the supply chain management function? One reason is financial, of course.

The supply chain is a major cost center ranging from 50 to 80 percent of the cost of sales (varies by industry). As a result, it is typically easier to reduce costs by a relatively small percentage and get the equivalent contribution to profit of increasing sales by a much larger percentage. For example, a company with a 10 percent profit margin and a supply chain cost of 60 percent of sales would need to increase sales by 4 dollars to have the same impact on the profit margin as a 1-dollar supply chain cost reduction, and we all know how hard it is to increase sales in the current economic environment.

Another reason is operational. There is something called the "bullwhip" or "whiplash" effect (see Fig. 1.2). Basically, it describes the magnified effect (especially on inventory, operational costs, and customer service) that occurs when orders move up the supply chain. This can be caused by a variety of things such as forecast errors, large lot sizes, long setups, panic ordering, variance in lead times, etc.

There is now, of course, more of a "need for speed" in the current society in general and specifically in business. The increased use of outsourcing, global supply chains, e-commerce, and shorter product life cycles has been a double-edged sword for supply chain professionals.

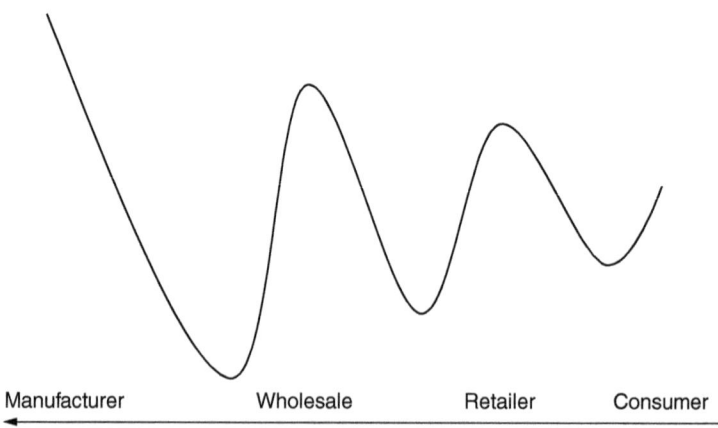

| Manufacturer | Wholesale | Retailer | Consumer |

Figure 1.2 Bullwhip effect.

On the one hand, it has given added exposure and importance to the function, but on the other hand, it has put added pressure on the supply chain to be efficient while it has become stretched and more complex (more of a supply "web" than a chain) at the same time.

Technology (explored in Chap. 12) can be an enabling partner on the Lean transformation journey, without which outsourcing, global supply chains, e-commerce, and shorter product life cycles wouldn't be fully attainable in the first place.

An efficient Lean supply chain can not only be used to improve the aforementioned financial and operational aspects of a business, but it can also be used as a competitive tool.

Kelly Thomas, vice president of manufacturing at JDA Software, Inc., believes that "a leaner, more operationally efficient supply chain translates directly into a company's financial returns. Companies can drive significantly higher gross margins, allowing them to invest much more in research and development (R&D) and sales and marketing. For example, Cisco typically runs the highest gross margins in the high tech networking equipment space, allowing it to outmaneuver its rivals in both research and development and marketing. This is a virtuous cycle, since this increased R&D spending typically results in better products, allowing the company to further its lead in the market."[Thomas, 2011]

The Lean Supply Chain Report by the Aberdeen Group, published in 2006, pointed out that there were a number of pressures that are driving Lean beyond the factory, including the need to improve operational performance and reduce operating costs, and customers demanding shorter order cycle times [www.aberdeen.com, 2011].

They point out that in some sectors, such as automotive, becoming Lean has become a requirement mandated by customers, while in general, in many industries, customers want shorter lead times, smaller order quantities, and lower prices. In order to do this, companies need to reduce their costs and make their supply chain operations more efficient.

In *Operations Management*, Roger Schroeder mentions two types of competitive supply chain strategies based on the type of product that is being supplied, imitative versus innovative.

Typically, imitative products have very predictive demand and therefore can use an efficient, low-cost supply chain strategy (e.g., generic over-the-counter drugs). Innovative products, on the other hand, have very unpredictable demand and require a faster, more flexible (and thus more costly) supply chain strategy [Schroeder et al., 2010].

It is not a one-size-fits-all type of thing, though, as an individual company can have multiple supply chains based upon their product lines. On the other hand, in *Operations Management*, Heizer and Render discuss using the supply chain to support three different competitive operations strategies [Heizer and Render, 2010]:

1. Differentiation—Where your product or service is better or different from the competition (e.g., Jet Blue Airways)
2. Cost Leadership—Low-cost strategy (e.g., Emerson Electronics)
3. Response—A quick response strategy (e.g., FedEx Corporation)

Again, a company may employ multiple strategies, which of course impact the supply chain used to support the strategies.

The Lean Supply Chain Report mentions top actions for Lean supply chains. Best-in-class companies are now focusing more on adding value from the customer perspective, while average and laggard companies are still focusing more on reducing non-value-added activities and reducing inventory and assets. All are also working on increasing supply chain flexibility and implementing a continuous improvement culture throughout the supply chain.

The report also points out that some of the top barriers to Lean strategies in the supply chain are cultural change required, top management commitment, and lack of participation of suppliers and other partners [www.aberdeen.com, 2011].

This is generally good news, but it shows that, in general, there is much work to be done (and many battles to be fought). So it is even more important to do it right early in the transformation process.

That is what this book is all about.

While this book will give the reader some background on the definition and history of Lean, it will also provide both the tools and practical applications to be successful.

The reader will also note that, as in many areas of business, there are lots of acronyms in this area of study, such as 5S-workplace organization (5S), total productive maintenance (TPM), and just-in-time (JIT), etc. which will be explained as we go along.

This book will describe how to identify and eliminate waste *specifically* in the supply chain and logistics function, types and examples of applications, and the tools to make it all happen. It includes snippets from articles and quotes from industry experts interspersed throughout the book to give the reader insight into current industry trends, as well as the use of technology to enable a Lean supply chain, internally as well as downstream and upstream.

CHAPTER 2

Historical Perspective: From Lean Manufacturing to Lean Enterprise . . . the Need for Speed

Evolution of Lean

In order to understand Lean and its current and future applications, it is important to first briefly review the history of manufacturing (Fig. 2.1) and how the concept of Lean originated and evolved. A good reason to do this is, as George Santayana once said, "Those who cannot remember the past are condemned to repeat it."

As mentioned previously, Lean is a team-based form of continuous improvement focused on the identification and elimination waste from the customer's perspective. Obviously, things haven't always been that way. If we look back the start of manufacturing hundreds of years ago, most goods were made by individual craftspeople or artisans.

Early concepts like *labor specialization* (Adam Smith), in which an individual was responsible for a single, repeatable activity, and *standardized parts* (Eli Whitney), helped to improve efficiency and quality. Up until that point, the individual craftsperson made most if not all of the product (furniture, wagons, etc.). If a wagon wheel broke, it had to be made from scratch and might not even be exactly the same as the wheel it replaced!

Around the turn of the 20th century, the era of scientific management came about, in which concepts such as *time and motion studies* (Frederick Taylor) and *Gantt charts* (Henry Gantt) allowed management to measure, analyze, and manage activities much more precisely.

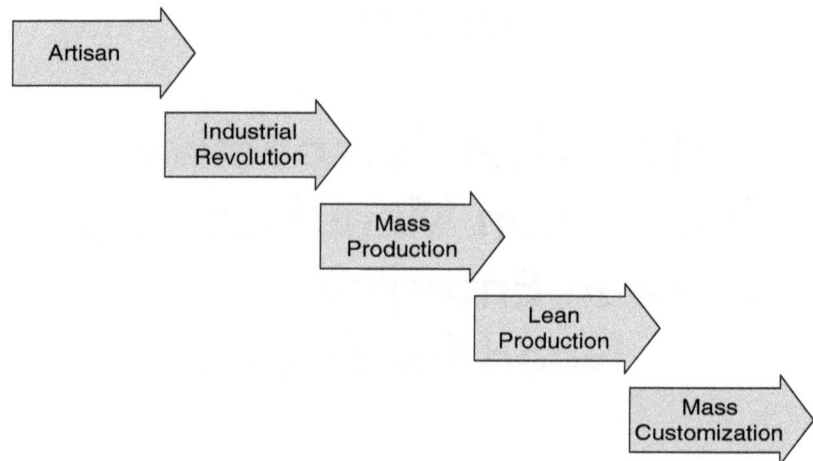

Figure 2.1 History of manufacturing.

The really big advancement came in the early 1900s with the era of mass production. Concepts like the moving assembly line (e.g., Ford Motor Company), *economies of scale* (produce large quantities of the same item to spread fixed costs), and statistical sampling were applied. Today, this concept is known as a *push* process, which is the antithesis of today's *demand pull* (by the customer) Lean thinking.

As the saying used to go, you could have any color Ford Model T, as long as it was black! In a push process, goods are produced in advance of actual demand and kept in inventory (typically based upon some kind of forecast) with the hope that customers will buy it.

In the 1980s, we started hearing about a new concept called *Just in Time* (JIT). This concept actually originated out of necessity by the Japanese after World War II when resources were scarce. It is a method of keeping minimal inventory of material (or information)—not too much and not too little. The demand for the JIT inventory is determined by downstream activities (ultimately the customers) that use it. A great tool to implement this concept is *kanban*, which will be discussed later in the book, but in general terms, *kanban* is a visual method for replenishing inventory that is withdrawn or consumed by a downstream process.

JIT, *kanbans*, and other concepts and tools, such as *total quality management* (TQM), *electronic data interchange* (EDI), and *employee empowerment* emerged. Many came from Japan, using statistical and other concepts taught by Americans after World War II such as W. Edwards Deming and Joseph M. Juran. The results were evident as Japanese products gradually went from being poorly made (e.g., cheap toys and transistor radios in the 1960s) to high performance and quality (e.g., Toyota, Nissan, etc., had a reputation for quality from the 1970s onward).

The Japanese also laid the groundwork for the concept of "demand pull" or Lean systems (sometimes referred to as *flow manufacturing*, *flexible manufacturing*, JIT, and other terms) to emerge. The true precursor to this is the much written about *Toyota Production System* (TPS), which was started in the late 1940s in Japan and which focuses on continuous improvement and respect for people. The objectives are to design out overburden (*muri*), inconsistency (*mura*), and waste (*muda*). TPS encourages employees to get to the source of a problem or issue by focusing on waste (which will be discussed in much detail in the next chapter).

The Need for Speed

This brings us to today and the future. While there is still an ongoing focus on cost and quality, there is now more of a "need for speed." This is a result of many converging technologies and cultural shifts. The advent of the Internet and e-commerce in the 1990s enabled news and information to travel at "warp" speed. That, combined with the "me too" generation's need to have things now, has both increased the need for speed to acquire goods and services, and also shortened product life cycles (e.g., cell phone models become old or obsolete in months not years).

The Internet when combined with massive enterprise resource planning (ERP) software systems (for which demand exploded as a result of the "year 2000" or Y2K technical issues) allowed manufacturers to become interconnected with both customers and suppliers to share and collaborate.

The world has become a global economy with companies sourcing product and material worldwide in search of the best quality at the lowest cost. E-commerce and ERP systems have lowered the boundaries for entry to the global economy for smaller companies as well, allowing them to compete anywhere, anytime against larger competitors.

All of that has led us on a path to mass customization, which is the capability to combine low per-unit costs of mass production with the flexibility associated with individual customization. One of the best examples of this in manufacturing is Dell Computers. They take your highly customized order for a computer (i.e., various combinations of monitors, hard drives, memory, etc.) and both assemble and ship within 24 hours. They can produce in small batch sizes and keep minimal inventory on hand. Dell works closely with suppliers (many of whom locate nearby) to continuously resupply inventory on a JIT basis. Instead of weeks and months of supply of inventory on-hand, Dell only keeps days and hours.

Dell is one example of a true demand-pull system, driven by end-customer demand. However, not all companies are candidates for this type of production. In fact, most companies that I've visited or know of use some kind of a combination of push and pull systems.

Lean Office

Around the year 2000, Lean manufacturing began to move from the shop floor to the office, as it became apparent that waste was everywhere and that offices shared some of the same characteristics of manufacturing, such as batching, setups, equipment failure, standardized work, etc. In fact, as much as 60 to 80 percent of a product or service lead time can be found in the office environment [which may include functions as diverse as customer service, order management, quoting, engineering, and research and development (R&D) to name a few].

The benefits of Lean Office vary, but include:

- ▲ More flexibility and responsiveness
- ▲ Reduced lead time
- ▲ Reduced errors and
- ▲ Extra processing
- ▲ Improved utilization of personnel
- ▲ Reduced transactions
- ▲ Simplified processes

Lean Supply Chain and Logistics Management

It has only been in the past 3 to 4 years that the concept of Lean has moved to the supply chain and logistics management environment. I believe there are a number of reasons for this.

A major reason is that, as previously mentioned, Lean started in manufacturing (especially repetitive, assembly-line manufacturing), then gradually moved to other manufacturing processes, such as continuous flow (e.g., chemical, food, and beverage) and, somewhat, to batch processing or job shop (smaller, often customer-specific production). Most manufacturers wanted to first "Lean out" within their "four walls" before working heavily with customers and suppliers. So, in a way, it is a natural evolution to move to the supply chain and logistics area. Companies now realize that they can only take things so far without collaborating and partnering more closely with customers and suppliers. Otherwise, in many cases, they just push their inefficiencies on to suppliers (e.g., JIT of raw materials) and are constantly frustrated with distorted and volatile customer demand (i.e., the bullwhip effect).

Additionally, while there has always been an emphasis on reducing costs, the recent economic meltdown has focused the red, hot light on supply chain management even more. So these days, any tool that can wring inefficiencies out of a system draws people to it.

Lean Six Sigma

There has also been a convergence of two different concepts, Lean and *Six Sigma*, referred to as *Lean Six Sigma*. In fact, many universities and organizations offer Lean Six Sigma certification programs.

According to the Bain & Company Web site (www.bain.com) "Lean Six Sigma is a blend of two concepts: Lean manufacturing, which is aimed at reducing waste, and Six Sigma, which helps companies reduce errors. Together they can help companies reap the benefits of faster processes with lower cost and higher quality."

However, Bain & Company has also learned through their surveys that "despite its growing popularity and impressive results at some companies, Lean Six Sigma often fails to deliver expected results. Our recent management survey of 183 companies found that 80 percent are not achieving their expected value from Lean Six Sigma efforts, and 74 percent are not gaining the expected competitive advantage because they have failed to achieve their savings targets." This is similar to things that have been written about Lean in general in the United States. Key success factors will be discussed in Chap. 10. [www.bain.com, 2011]

The concept of Six Sigma was originated by Motorola in the early 1980s and is now used in many industries. The term Six Sigma refers to a process that has 99.99966 percent of products produced free of defects (statistically speaking).

While, as mentioned previously, Lean is a team-based form of continuous improvement which uses relatively simple concepts to make improvements and covers the *entire* process or value stream, starting from the customer end working its way upstream to suppliers, Six Sigma is a tool (heavily statistical) that looks at individual steps in the process and attempts to identify and remove defects and variability. In general, Lean tries to reduce waste in the production process, and Six Sigma tries to add value to the production process.

Cycle Time versus Processing Time

While Lean is an easy concept to understand, it requires a slightly different way of thinking for everyone involved. First, it is important to understand the difference between cycle or lead time and processing time (Fig. 2.2).

Cycle time refers to the time required to finish an operation. An example of this would be the time it takes to manufacture and ship a widget (e.g., 5 days). This is the time it takes from when a customer places an order to when it is shipped (representing the cycle time, which of course varies by industry). That cycle may include manufacturing functions such as cutting, drilling, polishing, and packaging, which may only take minutes to an hour. These activities, where raw materials or inputs are transformed to finished product or outputs are referred to as *processing time* and add

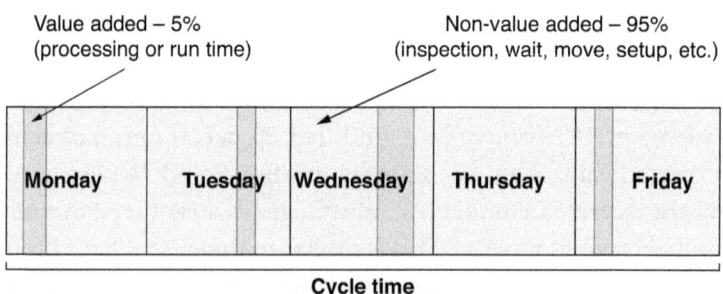

Value added – 5%
(processing or run time)

Non-value added – 95%
(inspection, wait, move, setup, etc.)

| Monday | Tuesday | Wednesday | Thursday | Friday |

Cycle time

Figure 2.2 Cycle versus processing time.

- **Value-added activities**
  - ✓ Activity that transforms material or information and that the customer is willing to pay for
- **Non-value-added necessary activities**
  - ✓ Activities that add no value but are required or necessary based upon regulatory, state of technology, etc. requirements
- **Non-value-added activities**
  - ✓ Activities that create no value in the eyes of the customer

Figure 2.3 Value-added versus non-value-added.

value from the viewpoint of the customer (i.e., what they are willing to pay for), but are typically a very small part of the process (5 to 10 percent).

From a Lean perspective, that cycle time is actually much longer than our widget example of 5 days, as there are many other *non-value-added* activities, such as product waiting for inspection, waiting to be moved, move time, setup time, and various administrative activities to name a few (see Fig. 2.3). As a result, inventory builds up to cover these non-value-added activities. It can take the form of raw materials, work in process (WIP) and finished goods inventory. When those inventories are converted to days of demand, the "true" cycle time, at least from a Lean perspective, balloons from 5 days to many weeks or months.

Historically, management has typically looked at increasing the speed of the processing time (e.g., increased hits/minute on a punch press, bottles filled/minute, etc.), while not looking as closely at reducing or eliminating non-value-added activities, as in many cases, they are hidden by inventory or just considered to be part of normal business operation.

It may make perfect sense to increase processing time, but in many cases it may not help the overall process efficiency, which Lean is all about.

Takt Time

There is a term called *takt time*, which refers to the production rate needed to meet customer demand for a family of products or services ("family" refers to an items or services that have mostly the same production or delivery steps). It is typically measured by dividing work time available by units required. In a Lean process, it is important that every step can meet this takt time. So in some cases, speeding up the processing time may actually be the wrong decision and create more non-value-added activities to the process.

Dock-to-Dock Time

One way to measure the effectiveness of a process is called *dock-to-dock time*, which looks at how long key material or product is in a facility. The longer it sits, the higher the cost and lower the throughput. It is also good to think of the order-to-cash cycle as a Lean measurement too. Eliminating non-value activities is a way to shorten the cycle time so as to speed up getting paid by the customer, which in today's challenging economic environment is a good thing.

The classic visual for this is a pie diagram showing a small piece, representing value-added activities (5 percent) and the much larger section (95 percent) representing non-value-added activities, much of which (60 percent +) is pure waste. The idea is that instead of expanding the pie by adding more people and other resources, focus on reducing the non-value-added activities and transferring those people and resources to value-added activities.

The net result, if done properly, is increased throughput with the same or less effort. It's really about people working smarter, not harder, using ideas that, for the most part, they come up with themselves.

In Lean terms, non-value-added activities are referred to as waste, which we will cover in the next chapter in some detail. It is a different way of thinking but can be a very effective way to find areas for improvement.

CHAPTER 3

The Eight Wastes: Waste Not, Want Not

What Is "Waste"?

Many people think they already know what "waste" in a process is as a result of always hearing about government waste. While there are some similarities, it is not really the same. In fact, waste in the way we mean is a different way of thinking to most people. The idea of thinking of a process in terms of value-added versus non-value-added or wasteful activities is easy to understand and definitely makes it easier to identify and eliminate waste.

The most common way to describe these non-value-added activities is by using the concept of the *seven wastes*, which we'll get to shortly. Some like to add another waste (behavioral) when discussing the topic. In general, waste can be defined as anything that does not add value to a process. Typically, when a product or information is being stored, inspected, delayed, is waiting in line, or is defective, it is not adding value and is 100 percent waste.

The original "seven wastes" came from the Toyota Production System (TPS), discussed earlier. The seven wastes include unnecessary *Transportation* or *movement*, *Inventory*, excess *Motion*, *Waiting*, *Overproduction*, *Overprocessing*, and *Defects or errors*. A good way to remember this is the acronym TIM WOOD. You can also add the eighth waste of *Behavior* (or underutilized employees) to this as it can sometimes be the biggest waste of all.

These wastes are applicable to any process, whether it is manufacturing, administration, or supply chain and logistics. What follows are descriptions of each of the wastes, as well as some examples.

The Eight Wastes

Inventory Waste

 It is best to go out of the Tim Wood order when reviewing the eight wastes, and start with inventory, as it is both the most visible and is actually an end result of the other wastes.

Inventory is a buffer between suppliers, manufacturers, and customers and is needed to compensate for lead times (e.g., in transportation, manufacturing, etc.) and variability in the system, such as forecast errors, late deliveries, setup times, scrap or rework, quality problems, and downtime.

In an office environment, inventory might refer to information, such as customer orders, or supplies, but is still just as important as it directly impacts the cycle time mentioned previously in Chap. 2.

There are four kinds of inventory:

1. **Raw materials**—Typically purchased materials and components.
2. **Work-in-process (WIP)**—The transformation process has started but has not yet been completed.
3. **Finished goods**—Finished, saleable products.
4. **Materials, repairs, and operations (MRO)**—Inventory for equipment spare parts and supplies.

All of these types of inventory cost money to maintain. This is called holding or carrying costs. These costs can range from 15 to 30 percent of the value of a product and include cost of capital (i.e., borrowing costs or opportunity cost lost if the money was invested elsewhere), taxes, storage, insurance, handling, labor, obsolescence, damage, and pilferage.

For example, if a business bought $100,000 of a raw material, had holding costs of 30 percent, and did not use the material for a year, it would have actually cost the business $130,000.

In reality, businesses need some inventory and typically have to balance the tradeoff between the cost of carrying inventory and customer service when determining how much.

There is the old analogy of a boat on the water, with the level of the water representing the amount of inventory and the jagged rocks below representing variability (Fig. 3.1).

Excess inventory is really a "symptom" of the problem. It is often said that the idea is to lower the water level until the "rocks" show above the

Figure 3.1 Need to reduce variability.

water. In fact, in many companies, the finance department will pass an "edict" to lower inventories by x percent by year end, which may expose the rocks, but can also create significant customer service problems. It is perhaps more conservative, but more effective to take the opposite approach. Instead, identify the sources of variability and then, using analytical tools (described in more detail in Chap. 6), such as the Pareto principle (also known as the "80/20" rule), root cause analysis, and the "Five whys" (keep asking "why" until you get to the root cause), reduce or eliminate the variability and *then* reduce the inventory levels.

Transportation or Movement Waste

This type of waste can include transporting, temporarily locating, filing, stocking, stacking, or moving materials, people, tools, or information.

Ideally, when material is received, it should only be touched once to put it away and another time to pull it for consumption. However, the reality is that it rarely happens this way. Material may be moved from one place to another on the floor, put on a storage rack, pulled to remove some material, then returned to a different rack, etc. All of this excess movement is wasteful. Companies are not only paying a forklift driver to move the material, but each time it is moved, damages may occur, and each time material is moved, inventory accuracy may be affected. When material is returned to a different spot, there is the risk of losing it and accidentally ordering more (yes, that does happen).

In many cases, transportation waste can be something very obvious that you just learned to live with, like the copier being too far from your desk, paper and staplers kept too far away from the copier, no signs identifying areas or departments, or just simply poor office layout.

That is why the "foundation" concepts of layout and visual workplace are so important, and these will be described in more detail in Chap. 5. When looking at layout, think of "flow." This is important whether we are talking about a manufacturing facility, office, or warehouse.

Motion Waste

 The concept of motion waste is best described by the idea of having things you use more often closer to you (and at waist level) and things you use less often further away and higher up. Any motion that does not add value to the product or service is wasteful.

The Lean concept of *point-of-use storage* is applicable here. It basically means having just enough material or information nearby, which can be replenished when needed from further away (a *kanban*, which will be discussed in Chap. 6, is an excellent visual tool for this type of replenishment).

Some examples of motion waste are looking for tools, excessive bending or reaching, and materials placed too far away.

When thinking about motion waste, the term *ergonomics* should come to mind. Ergonomics is the science of how humans interact with equipment and the workplace. So in terms of motion, you don't just want to consider efficiency, but safety as well (i.e., avoiding back injuries, carpal tunnel syndrome, etc.).

Waiting Waste

The waste of waiting is simply time spent waiting on materials, supplies, information, and people that are needed to finish a task. Everyone, whether on the shop floor, in a warehouse, or in an office can easily identify with this type of waste. It is both frustrating and counterproductive.

In most processes, a great deal of a product's or service's lead time is spent on waiting. In many cases, the waiting is caused by the next operation. This can be a result of long setup times, large batch sizes, and downtime. The result can be larger than needed amounts of WIP inventory.

In an office environment, time can be spent waiting on equipment to start up, printer or computer breakdown, signatures, employees on different work schedules, and even meeting attendees not showing up on time (which never happens, of course).

In many warehouses or distribution centers, products can sit "waiting" between different steps in the process (e.g., receiving, putting away, replenishing, picking, packing, and shipping).

Overproduction Waste

Overproduction, and its sinister sibling, overprocurement, is manufacturing, ordering, or processing something before it is actually needed. This typically results in an excess of another major waste already mentioned, inventory (raw, WIP, finished goods, and MRO). In addition, this can result in longer than necessary lead times, higher storage costs, and potentially a greater amount of defects (which may be harder to detect) because of larger-than-needed batch size.

Overproduction inhibits the smooth flow of materials, a basic tenet of Lean. Instead of Just in Time (JIT, discussed in Chap. 2), it ends up being Just in Case!

In the office environment, this may involve preparing or printing paperwork earlier in batches as a result of long setup times (yes, there are setups in the office as well as the shop floor!), preparing a report early, or in its entirety instead of online or as an exception report, and memos and e-mails that copy "the world."

The warehouse may suffer from some of the office-type overproduction wastes, as well as others like pulling orders earlier than needed or ordering supplies and packaging materials in large batch sizes, for example.

Overprocessing Waste

Overprocessing happens when too much time or effort is put into processing material or information that is not viewed as adding value to the customer. This can also include using equipment that may be more expensive, complicated, or precise than is actually needed to perform the operation.

This may occur when there are unclear customer specifications, a product or service is continually refined beyond what the customer wants or needs, or a lengthy approval process is involved.

Examples of this can be overpackaging (ever open a Christmas toy for a child and wonder if all of the packaging materials were really necessary and what a "waste" they are from an environmental perspective?) and overchecking.

In the office, overprocessing can include things like sending the same information in multiple formats (fax, e-mail, and overnight delivery), repeating the same information on different forms, reentering data, and unnecessary information on a form.

Defect or Error Waste

In manufacturing, the waste of defects primarily refers to repairing, reworking, or scrapping materials. The further along that a defect gets, the more costly it is to the company as they may need to rework it into the system, scrap it and make it all over again, and in the worst case, have it returned from the customer (which can include safety and liability issues as in the Tylenol and Toyota recalls, for example). A lot of extra, non-value-added activities take place as a result, such as quarantining, reinspection, and rescheduling, possibly resulting in overtime and, ultimately, lost capacity.

In the office and warehouse, this can be errors such as those made during data entry, receiving, and picking and shipping the wrong product (or the right product, but to the wrong customer). This can be the result of a lack of standardized work and a lack of a visual workplace (to be discussed later, but can include no checklists, forms, or directions), poor lighting, and lack of training.

There are many causes for this type of defect, such as poor processes, too much variations, supply issues, insufficient or improper training, tools and equipment not properly calibrated or precise, bad layouts, excessive or unnecessary handling, and inventory levels that are too high (e.g., sits around longer so more potential for damage).

Behavioral Waste (or Underutilized Employees)

 Some add an eighth waste of behavior. This is critical to consider, as you need employee creativity and participation to eliminate the other seven wastes. However, in some companies, there is a culture of not wanting to question things, not taking risk, or not rocking the boat. You might hear someone say, "This is how I was shown how to do it," or "We've been doing things this way for years." If you're going to have a successful Lean journey, this type of behavior is unacceptable and must be changed. A company's culture will be discussed later in more detail, but suffice it to say, a culture of team-based continuous improvement is a must.

You must fully utilize and leverage employee knowledge and skills, and offer proper training and opportunities for advancement to guarantee success.

Thinking Differently

As you can see, the concept of waste in Lean thinking is not really that complex. It is really just a different and easy way of looking at things. Once you start thinking this way as a team, it becomes easier to see where waste lies in your business. The next step is to take a step back and see where some of these wastes might exist in your supply chain.

CHAPTER 4

Lean Opportunities in Supply Chain and Logistics: Forest for the Trees

Sometimes we are all so busy "fighting fires" that we can't see the inefficiencies in our own system that may cause the fires in the first place. Lean teaches us to take a step back, take a deep breath, and get to the root cause of the fires.

For the purposes of this book, the SCOR (Supply Chain Operations Reference) model discussed in Chap. 1 works best to organize our thoughts on the subject (i.e., Plan, Source, Make, Deliver, and Return). We will look at some of the areas of waste in the supply chain and logistics management function in this chapter and then in Chaps. 5 and 6 describe some of the tools that can be used in the battle against waste in the supply chain.

Plan

A good plan must include a good understanding of current and future demand. In many businesses, forecasts drive the entire business. Forecasting is actually a mix of art and science, and many companies have spent tens of millions of dollars on expensive, complex systems, and then later blamed the system itself for bad forecasts. The classic example as described in "Beware the Promises of Forecasting Systems" by Ben Worthen [www.cio.com, 2003] is the Nike disaster in 2000, in which the company ended up writing off $400 million of inventory because the system had been so inaccurate. The "state of the art" forecasting system did not communicate well with Nike's existing systems, and in fact, some of the data had to be entered into the new system by hand, which increased the chance for data entry errors. The system was used as a kind of "black box," generating automated forecast projections, which had high rates of error. Nike ended up ordering $90 million worth of shoes that were very poor sellers, as well as having an estimated $80 to $100 million shortfall on its more popular models.

As a result, Nike's stock plummeted. Could this have been avoided? Of course. Many people like to just "press a button" and get results. The problem with having a black box that automatically generates forecasts is that no human input is involved.

Most, if not all, best-in-class companies know that the forecasting process is collaborative in nature. All forecasts are wrong—it's just a question of *how* wrong. Targeting and minimizing variability is key. A good collaborative forecasting process typically uses solid but straightforward models like time series and linear regression to give a decent baseline statistical forecast, which then is reviewed by the planner (on an exception basis) and aggregated (in various forms such as family, class, etc. and different units of measure) and then shared and reviewed with different parts of the organization for further enhancement (e.g., Fig. 4.1). Feedback, including point-of-sales (POS) data and forecasts from customers, is critical in this process as well (and will be discussed later in Chap. 13, when thinking outside the "four walls" and discussing *collaborative planning, forecasting, and replenishment* or CPFR). Accurate and timely feedback can minimize or avoid some of the bullwhip effects mentioned in Chap. 1.

It is also critical to set and measure forecast accuracy targets. As a result, it can be helpful set accuracy targets by ABC code. This uses the Pareto principle (or 80/20 rule), which states that a small number of items typically generate a large percentage of sales and/or profits (e.g., Burger King's Whopper, fries, and Coke). The A items, as they are the biggest sellers, typically have a tighter band of forecast accuracy (target and actual), but also require more time spent in terms of developing forecasts as they are so important to the company. They have relatively small days of supply of inventory as a result of their high volume and inherent forecast accuracy, as well as the fact that they are manufactured or ordered more frequently. C items (and to a lesser degree, "B" items) typically sell in small amounts and are more volatile, so they require wider forecast bands of accuracy and less time spent on them (to partially make up for this, a company may decide to keep many days of supply of inventory, which might not really amount to much anyway as they are small sellers).

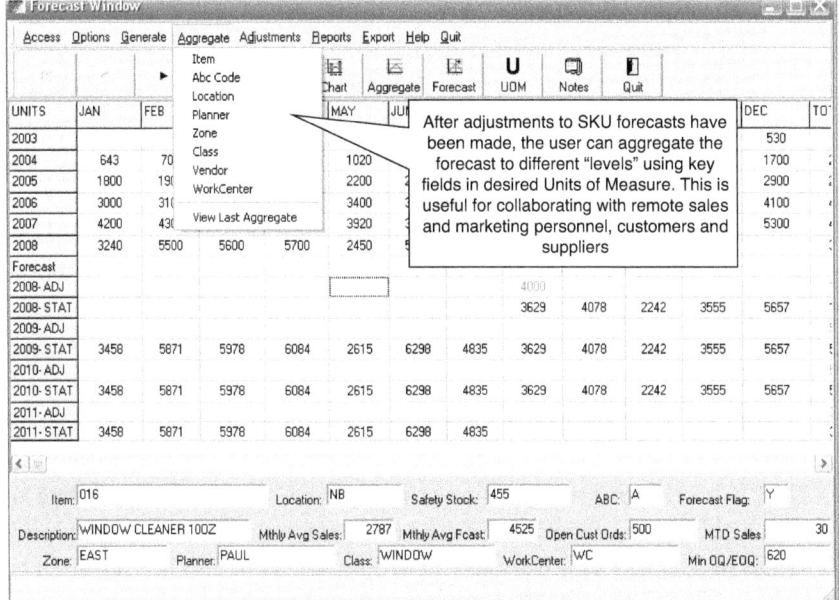

Source: PSI Planner™ for Windows®, Logistics Planning Associates, LLC, Copyright 1997–2010.

Figure 4.1   Forecasting screen.

Wastes in Forecasting

Some wastes that can occur in the forecasting process are found in both the process itself, and as a result of the process (i.e., bad forecasts). They include:

▲ **Letting the system run on its own**—A good example of this is HAL from the movie *2001: A Space Odyssey*. As mentioned in the Nike case, this can result in disaster. Human training, collaboration, and intervention are critical.

▲ **Using budgeted forecasts for operations**—Many companies use budgeted forecasts (usually from sales and/or marketing) for forecasting and do not even bother to update them during the year. The obvious problem with this is that:

▼ The forecasts were developed manually and from a subjective viewpoint. A good rule of thumb in this case is to take forecasts supplied by the sales group and divide them in half, which may give you a reasonable forecast.

▼ They quickly become "stale" and inaccurate if not updated regularly. Some companies even try to operate under a "two number" forecasting system, in which functions such as sales and marketing use a budgeted number while operations uses a statistical forecast. Then everyone wonders why things are out of control. Needless to say, this can become a total disaster.

▲ **Adjusting forecasts to compensate for supply issues**—People have a natural tendency to want to adjust a forecast because of potential supply issues. It is important to first come up with your "best guess" forecast and address those issues when discussing potential supply issues.

▲ **Using sales or shipment data versus demand or order data**—This used to be more of a problem because of the cost of computer memory, but it is critical to the accuracy of a forecast to use historical data of what the customer ordered, when they ordered it, and in the quantity they ordered. If you use sales or shipment data, you are doomed to repeat problems of the past (e.g., the order shipped short, late, or from an alternate location).

▲ **Aggregate versus SKU (stock keeping units or an item at a stock keeping location) accuracy**—It is a "given" that forecasts are more accurate at the aggregate level than at the SKU level. While, as already mentioned, it is useful to look at higher-level forecasts, accuracy measurements and control at all levels is important (variance targets of course may be greater at the SKU level than the aggregate level).

▲ **Poor communication**—Lack of communication or a "silo" or functional mentality can result in inaccurate forecasts and create waste through the entire process, which could have been avoided. For example, if someone in sales knows about a new customer and does not pass on some information (with sales estimates), operations will end up scrambling at the last minute to meet the order, if not ending up missing the opportunity altogether.

▲ **Not seeing the forest for the trees**—As mentioned in this chapter's title, some people just become used to fire fighting and are so deep into the detail they cannot see the big picture. A good forecasting process enables everyone to plan, not react, reducing the resultant wastes that are created by fire fighting. This is part of a sales and operations planning(S&OP) process which will be discussed in detail in Chap. 10.

Source

Supply chain costs can range from 50 to 80 percent of a company's sales depending on the industry. Therefore, it is not difficult to see why it is an area of interest in terms of looking for waste.

Lean sourcing or *procurement* is a different way of looking at and working with suppliers. There is a greater use of partnerships and alliances as well as a greater need for coordination and collaboration.

Traditional supply chains are managed more on a cost basis, negotiating with many suppliers. While this may still be effective in some instances (e.g., commodities), Lean procurement is all about long-term partnering with fewer, longer-term suppliers with less reliance on low-cost bidding. Motorola, for example, has eliminated traditional supplier bidding by adding emphasis on quality and reliability and in some cases may sign contracts that are in place throughout a product's life cycle [Heizer and Render, 2011]. In this way, the relationship can be mutually beneficial. The value is created by economies of scale and long-term improvements (see Table 4.1).

As a result of this type of relationship, where trust is very important, suppliers are more willing to get involved in JIT partnerships and share in the design process and be willing to contribute technological expertise. For example, when Cessna Aircraft opened a new plant in Kansas, they set up consignment and vendor-managed inventory programs with some select suppliers. One supplier, Honeywell, was allowed to maintain avionic parts onsite. Other vendors who participated kept parts at a nearby warehouse to supply the production line on a daily basis. This was a win-win situation, as Cessna was able to execute JIT inventory replenishment for parts, and their suppliers gained better insight into Cessna's production requirements and could offer suggestions for product improvements, thereby strengthening the relationship [Heizer and Render, 2011].

Some suppliers may be somewhat hesitant because of issues such as having too much reliance on one customer, shorter lead times, smaller order quantities, etc. As a true partnership, the customer must be willing to work with the supplier and share costs, training, and expertise so that they are not just passing off their problems upstream. Of course, you need to always have a "backup" plan and only single source (i.e., one supplier for an item) where there is very little risk involved (e.g., commodity-type item, easily substituted part, etc.).

Table 4.1 Lean Supply Chain Characteristics

Characteristic	Traditional Supply Chain	Lean Supply Chain
Suppliers	Many	Few
Interactions	Confrontational	Collaborative
Relationship focus	Transactional	Long-term
Primary selection criteria	Price	Performance
Length of contract	Short-term	Long-term
Future pricing	Increased	Decreased
Lead times	Long	Short
Order quantities	Large lots	Small lots
Quality	Extensive inspection	Quality at the source
Inventory (supplier and customer)	Large	Minimal
Information flow	One way	Two way
Flexibility	Low	High
Product development role	Small	Large (collaborative)
Trust	Limited	Extensive

There are many Lean opportunities in procurement including:

▲ JIT, such as in the Cessna example above. There may also be a potential application for *vendor-managed inventory* (VMI), in which a supplier manages its customer's inventory of parts and supplies (to be discussed in detail in Chap. 6).

▲ Batch size and lead time reduction—producing smaller quantities of items more frequently, thus reducing inventory and cycle time.

▲ Blanket orders—in which a customer places a single purchase order with its supplier containing multiple delivery dates scheduled over a period of time, in many cases at predetermined prices.

As they say, "If you can't measure it, you can't improve it." This applies to all of the applications mentioned in this book. By performing Lean assessments and supplier reviews, you can determine how lean your supplier is and what progress has been made toward that goal.

Make

Some might argue as to whether or not the Make process is part of supply chain and logistics management or just supported by it. Either way, there is no doubt that if not functioning properly, it can cause massive wastes in the *entire* supply chain.

Once you are comfortable with the demand side (i.e., sales forecasts), it is critical to execute an accurate, deliverable supply plan. This includes production planning for finished goods, as well as procurement planning for any materials (raw, component, or finished) sourced from a supplier.

In *Successful Lean Planning*, an article that appeared in the May/June 2010 *APICS Magazine*, Preston McCreary writes, "Planning is a key part of any manufacturing business strategy and possibly even more important in a lean-flow company. Planning determines how, when, and where a product is produced. Successful planning ensures customers get products on time and within cost expectations—creating more opportunities to sell. Poor planning will cost the company money, create chaos…and lead customers to buy product elsewhere. Lean planning uses the concept of pull versus push" [McCreary, 2010].

Rarely have companies been able to go completely from a push-type system, producing in large batches based upon a forecast, resulting in large amounts of finished goods inventory, to a customer-demand pull system, in which customer demand triggers activities such as distribution and production upstream. Most companies are somewhere in between. The goal is to get closer to "make what you sell" and have an ultimate (yet perhaps even unrealistic) goal of one-piece flow.

Make to Order (MTO) versus Make to Stock (MTS)

For many companies, it may be impractical to have a *Make-to-Order* (MTO) process, which is most conducive to one-piece flow and a totally pull system. As a result, they operate primarily under a *Make-to-Stock* (MTS) process.

The concept of *postponement* can also be a useful tool to bridge the gap between MTO and MTS. Postponement occurs when decisions about the transportation or transformation of product form in the supply chain are postponed until an order is received from the customer. An example of postponement would be when a computer manufacturer delays assembling the final customer order until it is released to the plant or warehouse for assembly or shipment. It is only at this point that the order is assembled and shipped as there may be an almost infinite combination of computers (and components), monitors, printers, etc. available. This reduces the number of items that a company needs to stock and keep track of.

However, even if a company operates in an MTS environment, customer demand can be used in combination with forecasts to drive more of a pull system, helping to eliminate some of the inherent waste found in the process.

Distribution Requirements Planning

In order to go from push to pull, a tool like *distribution requirement planning* (DRP) helps in the transition (see Fig. 4.2). The use of time-phased demand planning enables you to evaluate demand closer to the customer [at your distribution centers, and in the case of *collaborative planning, forecasting, and replenishment* (CPFR, discussed in Chap. 13), at your customer distribution centers]. This phased demand factors in consumption of the forecast by live orders, lead times, order minimums and multiples, scheduled orders (production work orders, purchase orders, and transfers), and scientific safety stock.

Many times when assembling production plans, companies neglect to include current demand information, as well as a "robust" safety stock to compensate for variability in demand and lead times.

It is critical that safety or buffer stock is scientifically calculated and *recalculated* on a regular basis. Typically, this number is based upon a desired service level by SKU, which calculates the number of standard deviations needed to cover that item's demand and lead time variability. It is always a good idea to calculate safety stock by ABC code with a higher service level (typically 99 percent) for As, lower for Bs, then lowest for Cs and beyond (some businesses have D and E items). By using this methodology, current demand and lead time variability is accounted for, as well as optimizing overall inventory levels.

In many cases, companies use a one-size-fits-all method for safety stock, which tends to bring overall inventory levels way up, while keeping service levels (especially for A and B items) lower than desired. The one-size-fits-all type of inventory planning usually results in having not enough of the "right" product and too much of the "wrong" product.

Step #1: Creating projected Ending Inventory

Current on-hand inventory is "netted" against the next planning period's Gross Requirements. Then any Scheduled Receipts in that same period are added in to create a projected Ending Inventory for that period.

In our example, the on-hand balance of 3,300 is "netted" against the greater of the next period's Forecast (206) or Open Customer Orders (0) plus any Dependent Demand* (0) to create Gross Requirements of 206. Then, the 2,300 units of Scheduled Receipts are added in to create a projected Ending Inventory of 5,394.

*requirements from other locations for this same item which are supplied by this location

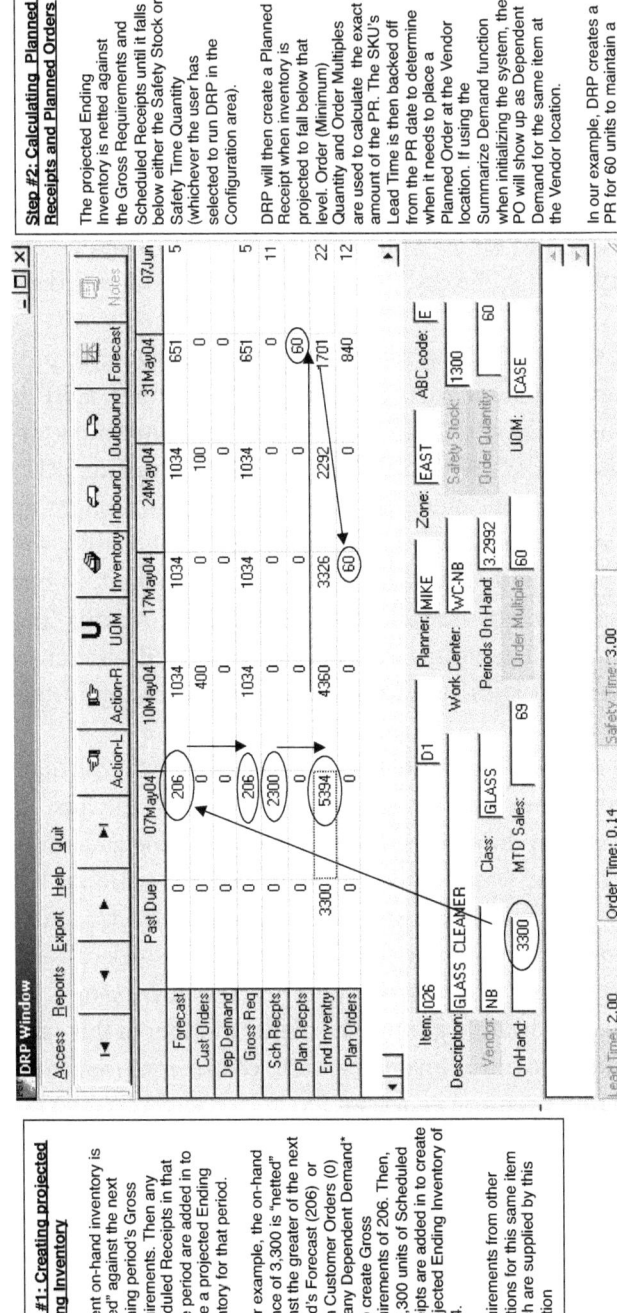

Step #2: Calculating Planned Receipts and Planned Orders

The projected Ending Inventory is netted against the Gross Requirements and Scheduled Receipts until it falls below either the Safety Stock or Safety Time Quantity (whichever the user has selected to run DRP in the Configuration area).

DRP will then create a Planned Receipt when inventory is projected to fall below that level. Order (Minimum) Quantity and Order Multiples are used to calculate the exact amount of the PR. The SKU's Lead Time is then backed off from the PR date to determine when it needs to place a Planned Order at the Vendor location. If using the Summarize Demand function when initializing the system, the PO will show up as Dependent Demand for the same item at the Vendor location.

In our example, DRP creates a PR for 60 units to maintain a Safety Time of 3 periods of future Gross Requirements (1,701). The Lead Time of 2 planning periods is backed off to determine when the replenishment needs to ship from the Vendor location

Figure 4.2 DRP screen and description.

Source: PSI Planner™ for Windows®, Logistics Planning Associates, LLC, Copyright 1997–2010.

35

This type of inventory model is used when demand is not constant or certain and the actual *reorder point* is calculated as

$$\text{Daily demand} \times \text{order lead time} + \text{safety stock}$$

Take, for example, the situation in which the average daily demand for a widget is 30 units, replenishment lead time is 3 days, and desired safety stock is 15 units. In this case, when the inventory gets down to 45 units, a replenishment order is placed for 30 units. The goal here is to always try to have 15 units of safety stock at the end of each planning period (i.e., days, weeks, or months).

While the use of safety stock is great for longer-term, aggregate production planning (and for item parts and supplies planning), DRP is more effective when using *safety time,* in which you have an inventory target of a specific number of days, weeks, and months of supply of an SKU's forecast. This uses a *reorder time* model, in which inventory is brought up to a target of days of supply based upon the SKU's forecast. It enables you to match an item's "peaks and valleys," which is very important, as one week of supply for an item may be 100 units during some months, and 1,000 units during others. Safety time should also be applied differently based upon SKU characteristics (e.g., A items carry lower safety time, C items much greater).

Even when using reorder time in DRP, it's a good idea to still calculate safety stock as a "second view" (or the greater of the two), at the least. If you find that there is a great difference between the safety time and safety stock requirements, you may want to investigate further. For example, if a nonseasonal item calls for a safety time of 2 weeks of supply, which equates to 100 units based upon the forecast and the calculated safety stock calls for 500 units, it may indicate that this item has a very erratic sales history. To be on the safe side, you may want to go with the higher number in this case.

By using a tool like this, much inventory waste can be removed from the system, while improving customer service levels.

Deliver

As flow and velocity are critical to Lean thinking, the Deliver process can contribute greatly to these goals. This includes both transportation and distribution operations.

In transportation, areas of Lean thinking may include:

⚠ Core carrier programs to reduce number of suppliers and develop collaborative long-term relationships

⚠ Improved transportation administrative processes and automated functions, such as *transportation management systems* (or TMS; see Chap. 12 for more details), which may help to optimize mode selection, right-sizing equipment, pool orders, and combine multistop truckloads

⚠ *Cross-docking*, in which incoming materials are quickly passed through a distribution center, typically within 24 hours, to outbound trucks for final delivery

⚠ Reviewing import/export transportation processes that are both complex and ripe for waste

⚠ Gaining control of inbound transportation and increasing use of backhauls to reduce costs and improve productivity

⚠ Reviewing freight auditing and payment processes that are often manually processed and errors may not be caught until post-payment audit if at all

Warehouse operations usually have high levels of activity with people and products always in motion. However, this type of action doesn't always result in real productivity.

Even though there appears to be constant motion in a warehouse, actual customer orders may move rather slowly through the system. Ideally, information and materials should "flow" through a facility. In reality, they tend to be batched and sit in queues in between processing steps (think of an inbox in an office, or some pallets sitting in an aisle). As a result, this ends up increasing the lead time and results in a less than optimal use of resources. This can often result in wasted space in a warehouse.

In warehouse operations, waste is found throughout the basic functions of receiving, putaway, replenishing, picking, packing, and loading. This waste can be a result of:

⚠ Defective products or errors, which create returns
⚠ Overproduction or overshipment of products
⚠ Excess inventories, which require additional space and reduce warehousing efficiency
⚠ Inventory accuracy issues

▲ Waiting or searching for tools and equipment, such as a forklift or a ladder
▲ Waiting for parts, materials, and information
▲ Excess motion and handling
▲ Inefficiencies and unnecessary processing steps
▲ Material handling steps and distances as well as blocked aisles
▲ Information processes and computer issues (system running slow, downtime, multiple screens to access, scanner issues, etc.)

Return

Returns or reverse logistics, as it is commonly known, is the return of product from a customer for a variety of reasons (defective, damaged, wrong item, servicing, did not want, etc.). As a result, this process is, for the most part, a waste, as it is caused primarily by defects and errors upstream from the customer.

By focusing on upstream returns, this activity can be substantially reduced. However, as there will always be some returns, it is important to look at the process for authorizing and handling returns from a Lean perspective as well.

In *Implementing Six Sigma Principles in Reverse Logistics,* part of the proceedings of the 2009 Annual Meeting of Collegiate Marketing Educators, Servos et al. point out that "movement, processing, and corrections are three areas of cost that are particularly well suited to...waste elimination... in reverse logistics." [www.a-cme.org, 2011]

Most companies have some kind of *return merchandise authorization* (RMA) process, whereby a customer must get some kind of approval to return a product before actually doing so (usually in the form of an RMA number provided by the supplier).

RMA processes usually involve several functions, many steps, and various people. Things to look at for waste may include:

▲ Reducing the number of RMAs that are open at any one time and setting limits for how long they can be open
▲ Reducing freight costs associated with returns
▲ Streamlining the actual RMA process trying to especially focus on wastes in movement, processing, and corrections
▲ Improving and enforcing compliance by both company reps and customers

▲ Eliminating return of unwanted parts (e.g., destroy in field if appropriate)

▲ Improving focus on spare/repair parts inventory

The SCOR model of Plan, Source, Make, Deliver, and Return is a great way to start to wrap your head around this complex subject and to identify general areas for improvement in the supply chain.

As the reader can see, there are *many* opportunities in supply chain and logistics management on which to focus and, in the process, eliminate waste. In the next chapter, we will discuss some of the basic Lean tools and examples of how to use them.

CHAPTER 5

Basic Lean Tools: You Can't Build a House without a Solid Foundation

The saying "you cannot build a house without a solid foundation" definitely applies when discussing Lean. In fact, there is something called "the House of Lean" (Fig. 5.1), which helps to illustrate this concept.

Although the importance of having a Lean culture as a key success factor has been discussed already (and will be discussed in more detail in Chap. 10), understanding how and when to use Lean tools will now be discussed in this chapter.

In many cases, *value stream mapping* (VSM; Fig. 5.2) is typically the next step taken by management after gaining a basic understanding of general Lean concepts. However, it is best to wait to address the details of VSM until Chap. 11, as you first need to understand the basic and advanced concepts and tools (in some detail) that can be used to deliver the opportunities for improvement identified in a value stream map.

Briefly, VSM is a mapping tool that is a 10,000-foot-level view of a process. Typically, it is for a family of goods or services from the customer working its way upstream all the way to key suppliers. A VSM is similar to a flowchart or process flow map, but one of the key differences is the "current state" map identifies value-added and non-value-added activities. The "future state" map, which can be thought of as a road map, attempts to reduce or eliminate the identified non-value-added activities.

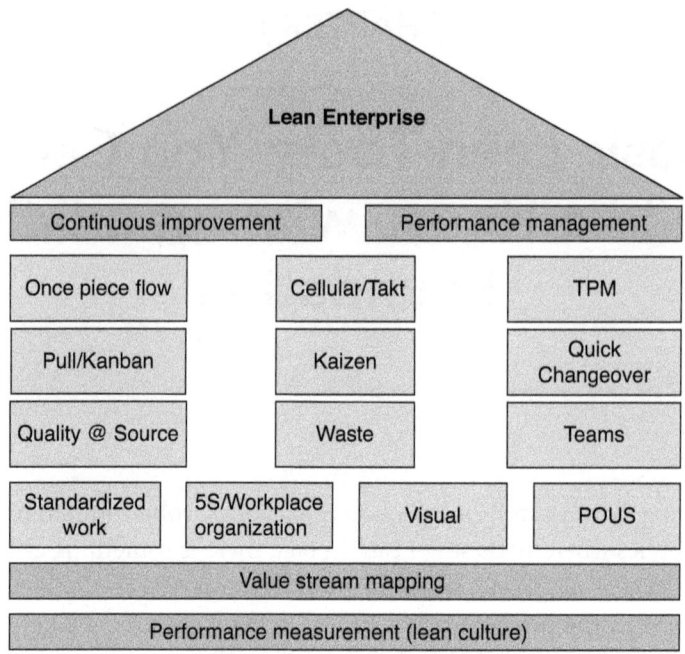

Figure 5.1 House of Lean.

Standardized Work

Standardized work refers to the standardization of best work practices—as the work is actually done in real life. The idea is to make work safe and repeatable with as little variation as possible along with high productivity. It is the best combination of employees, equipment, materials, and procedures.

There are examples everywhere of standardized work, including orders, drawings, *standard operating procedures* (SOPs), etc. In fact, standardized work is one of the foundation principles in the Toyota Production System (TPS).

We know in real life, while there may be SOPs in a binder on a shelf somewhere, most people do a job the way they were trained (or in many cases, how they learned it on their own). In many cases, this may not be the best way (i.e., in terms of method, sequence, etc.) but may be the way they were shown (e.g., "The guy I replaced showed me how to do this before he left.") or have done it for many years (e.g., "I've been doing it this way for years and it works for me."). The problem with this is that if everyone performs a task slightly differently, there may be variation which can result in waste.

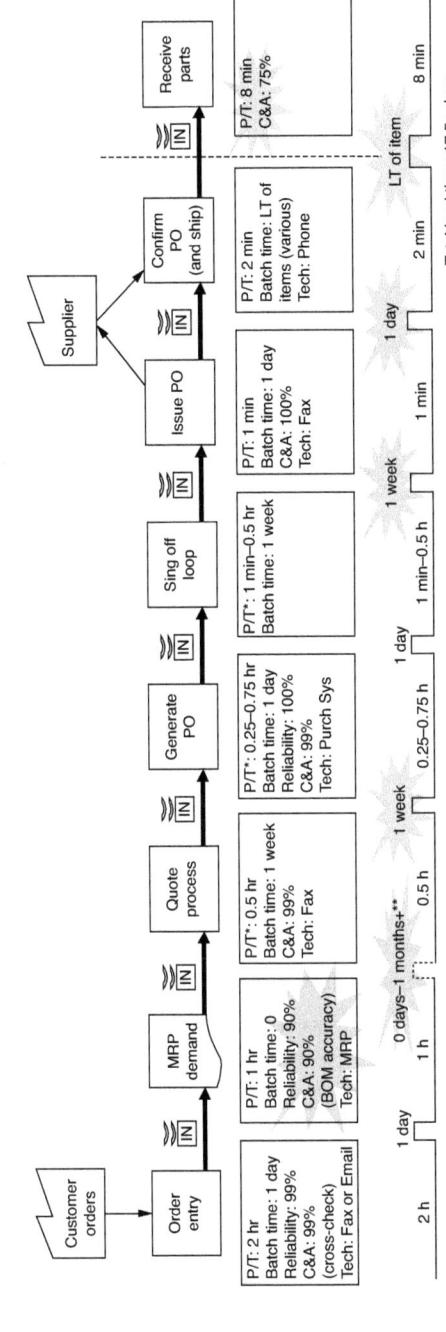

Figure 5.2 Current state value stream map example.

43

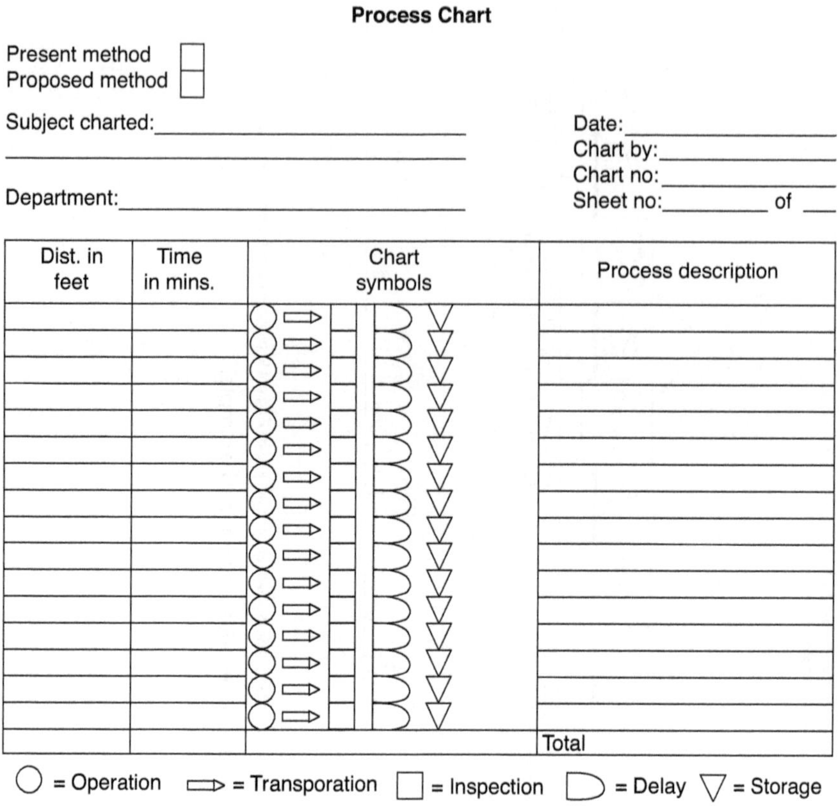

Process Chart

Present method ☐
Proposed method ☐

Subject charted:_____

Department:_____

Date:_____
Chart by:_____
Chart no:_____
Sheet no:_____ of ____

Dist. in feet	Time in mins.	Chart symbols	Process description
		Total	

◯ = Operation   ⇨ = Transporation   ☐ = Inspection   ◗ = Delay   ▽ = Storage

Figure 5.3   Process chart.

Usually it is best to get a team together of employees who actually do the work along with coworkers from other areas, document the steps in the process (using digital photography), and come up with an agreed-upon best practice minimizing waste in the process. It can be useful to use a tool such as a *process chart* (Fig. 5.3) to identify opportunities in the work process by capturing data for each activity, such as time and distance.

Visual Job Aids

It is then important to make this standard work more of a "visual job aid" (see Fig. 5.4) that is easy to understand and follow.

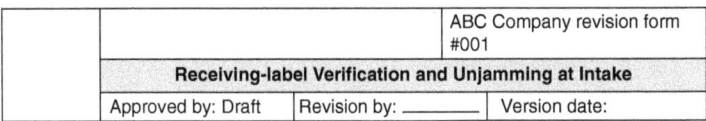

	ABC Company revision form #001		
	Receiving-label Verification and Unjamming at Intake		
	Approved by: Draft	Revision by: ⎯⎯⎯⎯	Version date:

Phase	Step	Comments	Visuals
1. 1st electric Sys-label verification	• Inspect LPN and carton label • Press reset button	• Numbers must match • If numbers match, if not, take carton off conveyer and see supervisor	
2. 2nd electric Sys-un-jam (clear)	• Un-jam ("clear") carton • Push reset button		

Figure 5.4   Visual job aid.

This type of visual job aid should then be placed in the area where the work is done (laminated for protection) so that it may be followed by everyone. It is always a good idea to consider language restrictions in the workplace when creating standardized work (e.g., English and Spanish versions).

Areas in supply chain and logistics that this can apply to are found everywhere. In the office and warehouse/distribution center are the most common and can include order processing, invoicing, and draw-ings. In the warehouse, pretty much all of the basic activities of receiv-ing, putting away, picking, packing, and loading can benefit from visual job aids (see Fig. 5.5).

Visual job aids are especially important in warehouse operations, both in the office and on the floor, as there is a large use of temporary workers, "lumpers" (outsourced workers who handle freight or cargo), and the use of *third-party logistics* organizations or 3PLs (i.e., outsourced logistics func-tions on a short-term or long-term basis).

Standardized work then leads to an organized workplace, which is neat, safe, and efficient, with a location for everything needed and the elimina-tion of anything that is not needed in the area.

		ABC Company shipping form #001
	Ship Load (non-con) Visual Job Aid	
	Approved by: Draft	Revision by: _____	Version date:

Phase	Step	Checks/Remarks	Visuals
1. Move pallet(s) to door	• Move 1–2 pallets from door staging lane to door for loading	• # of pallets depends on equipment used.	
2. Scan pallet(s)	• Scan pallet label with RF gun	• Select option #5 on RF gun ("Load Carton/Pallet"); then enter Store # and Door #, then scan "Manifest Label"	
3. Load cartons into trailer	• Load cartons into nose of trailer if possible	• The objective is to pack cartons in tightly and securely as possible. • Large, heavy cartons should go on the floor with arrows up.	
4. Return empty pallet	• Return to designated pallet return area for that door.	• Located by door staging lane	

Figure 5.5 Visual job aid (loading a truck).

Visual Workplace

Think of the chaos that would result in our everyday lives if there were no speed limit signs, no lines on the highway or in a parking lot.

In manufacturing terms, there is something called a "visual factory." These simple visual signals give operators the information to make the right decision. They are efficient, self-regulating, and worker-managed. Examples include visual job aids, signs, lines on the floor designating storage areas, aisles, work areas, etc., "andon" lights (i.e., "and on" a red light there is a problem, etc.), labels (color-coded in some cases), and *kanbans* (visual signals to replenish inventory as a result of downstream demand; e.g., Fig. 5.6).

The visual workplace is one of the fundamental concepts of Lean. This translates easily to the supply chain and logistics function. At first glance, a warehouse looks fairly organized with bar codes and labels on pallet bins, safety lines on floors, etc. However, a closer look usually reveals areas of clutter and disorganization (examples might include the supply and maintenance areas).

When implementing visual systems it is important that they are easily found where needed, easy and quick to understand, and provide meaningful feedback.

Figure 5.6   Simple kanban with visual signal example.

Layout

Another key concept in Lean is flow. By eliminating waste in a process, items keep flowing, as opposed to waiting in a queue, in an aisle, etc. Critical to this is the layout of a facility itself.

Typically, companies grow "organically" and put things where they fit, not necessarily where they best belong. There tend to be "monuments," which are big, heavy pieces of equipment which are difficult to relocate. This may not be conducive to the continuous flow of materials or information.

In a warehouse, as travel time is critical to productivity, good layout is essential. The idea of having your fast-moving or A items located closer to shipping (and down low) and your slower-moving C items farther away and higher up (commonly called "velocity slotting") is not always the case. Also it is important to have tools, equipment, supplies, and packaging materials always available and close to where you need them.

The same goes for the office where we tend to lose sight of how much walking we do to process an order, for example (it can add up to hundreds of miles per year of unnecessary walking!).

Good layout results in:

▲ Higher utilization of space, equipment, and people
▲ Improved flow of information, materials, or people
▲ Improved employee morale
▲ Improved customer/client interface
▲ Flexibility

5S: Workplace Organization and Standardization

5S, which stands for sort out, set in order, shine, standardize, and sustain, is a tool that results in a well-organized workplace complete with visual controls, improved layout, and order. It is an environment that has "a place for everything and everything in its place, when you need it."

5S produces a workplace that is clean, uncluttered, safe, and organized. People become empowered, engaged, and excited.

A workplace that is clean, organized, orderly, safe, efficient, and pleasant results in:

- Fewer accidents
- Improved efficiency
- Reduced searching time
- Reduced contamination
- Visual workplace control
- A foundation for all other improvement activities

In the supply chain and logistics function, especially in the case of warehouse operations, it is often the first place that Lean is implemented. The main reasons are that it is a good foundation concept for future improvements, and it is simple to understand and implement.

One of the leaders in this is Menlo Logistics, a Division of Con-Way (www.con-way.com). They dedicate an entire section of their Web site to Lean logistics and state that "Menlo Worldwide Logistics practices lean logistics to deliver superior supply chain performance and give its customers a competitive advantage. Lean logistics emphasizes minimization of all resources used in supply chain management. The lean logistics methodology uses proven lean practices and principles to reduce waste, complexity and error… adherence to 5S leads to better quality service, lower costs, higher availability, higher customer satisfaction, and more reliable deliveries." [www.con-way.com, 2011]

Worker productivity in a warehouse or distribution center is especially critical (e.g., cases per hour or CPH is a typical productivity measure), and 5S can be especially useful in this regard.

As was previously mentioned, the actual 5Ss stand for: sort out, set in order, shine, standardize, and sustain. They will be defined shortly, but before starting on a 5S *kaizen or improvement* event, one must first perform a workplace scan. Typically, this entails the following steps:

1. **Area map**—Usually drawn by hand, it should show the area being 5S'd, including all machines and materials located in the area. Colored lines should show the movement of materials (and information) in and out of the area. This is called a "spaghetti" map for a reason—by the end, it usually looks like a bowl of spaghetti! It is a good way for the group to understand where there may be opportunities for improvement and where things are not flowing.

2. **5S Audit**—There are many 5S audits available on the Internet. Basically, the facilitator leads the group through a "before" rating of the area in terms of each of the 5Ss (we will get there shortly), usually on a scale of 1 to 5, with 5 being the best. Typically, the first audit results in a fairly low score. The idea is that subsequent audits will find better results and can gradually be done less often.

3. **"Before" pictures**—It is always important to remember what the area looked like before it was 5S'd. The pictures can possibly be posted later on a 5S board with "after" pictures.

Once the team has selected an area and performed an initial workplace scan, the 5S process can start.

Sort Out

The first S is for sort out. This involves removing anything that is not needed in the area. Items can range from garbage, which can be disposed of immediately, to excess inventory, equipment, tools, furniture, etc. that don't belong there. As the saying goes, "When in doubt, toss it out."

Why do you need to do this? Throughput is increased as a result of improved work flow, communication between workers is improved, product quality is increased, wasted space is reduced, time spent looking for parts or tools is reduced, and overstocking is avoided.

Included in this phase is something called a *red tag* strategy. Simply put, a nearby area needs to be designated where sorted out items can be taken for future disposition. This is called a red tag area. Each item taken there should have a red tag attached with a variety of information, such as an assigned number, description, and recommended action (see Fig. 5.7).

Each red tag item should be on a corresponding red tag disposition sheet, which is communicated to the appropriate parties noting that red tag items must be claimed and removed within a certain time period (usually no more than 2 weeks), or they will be disposed of.

Surprisingly, there are quite a few categories and actions that can be taken during this process as can be seen in the sample disposition list in Fig. 5.8.

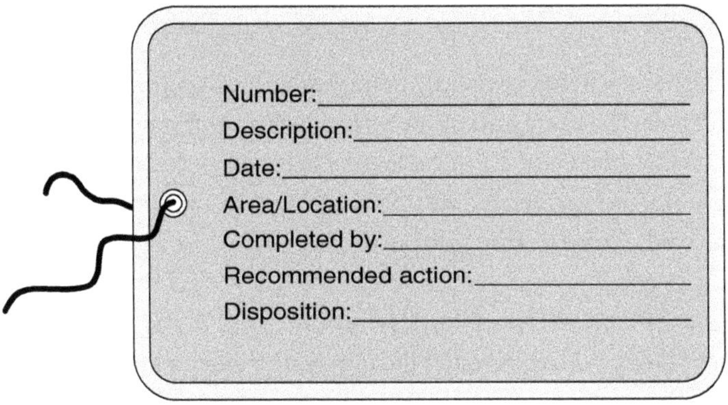

Number:_____
Description:_____
Date:_____
Area/Location:_____
Completed by:_____
Recommended action:_____
Disposition:_____

Figure 5.7 Sample red tag.

Category	Action
Obsolete	▲ Sell
	▲ Hold for depreciation
	▲ Give away
	▲ Throw away
Defective	▲ Return to supplier
	▲ Recycle
Scrap	▲ Remove from area to proper location
Trash	▲ Throw away
	▲ Recycle
Unneeded in this area	▲ Remove from area to proper location
Used daily	▲ Carry with you
	▲ Keep at place of use
Used weekly	▲ Store in area
Used monthly or less	▲ Store where accessible in facility
Seldom used	▲ Store in distant place
	▲ Sell
	▲ Give away
	▲ Throw away
Use unknown	▲ Determine use
	▲ Remove from area to proper location

Figure 5.8 Disposition list.

Set in Order

After unnecessary items have been removed, everything remaining should be set in order. As the saying goes, "A place for everything and everything in its place." During this phase, a great deal of thought should be given to the area's layout and flow of materials and information. It is a great time for the use of visuals as well. Color-coding and outlining are a few ways to make it clear where things go. This is a good time to have a label maker and masking tape, to temporarily label shelves, bins, drawers, etc. and mark floors. A few weeks later, after "living with it" for awhile, everything will be made permanent.

A great visual to create in this phase is something called a shadow board. It can be as simple as taking an ordinary peg board and outlining and describing the tools that go on it. That way, when workers see an empty spot on the board, they know that a tool is missing. It is also a good idea to label the tools hung on the board with the same information, so when someone finds a tool they know where it goes.

Now is also the time to consider excess motion waste. You can also consider what was found in the area map done earlier.

Shine

The third S is for shine. During this phase, everything is cleaned and sometimes even painted. One of the key purposes of cleaning is to keep all equipment in top condition so that it is always ready to be used.

If the shine phase is not done, problems that can come up including poor employee morale, safety hazards, equipment breakdowns, and even possibly an increased number of product defects.

During the shine process, we clean away trash, filth, dust, and other foreign matter. Contamination can include debris, oil, documents, water, dirt and dust, food and drink, poor work habits, and materials left by other people, such as maintenance.

As a group, you need to determine what needs to be cleaned, who is responsible, how it is done, and what tools are needed. Cleaning supplies should be neat, clean, well-organized, and readily available (see Fig. 5.9). In many cases, areas have a "5-minute cleanup" at the end of a shift. Housekeeping checklists are always a good idea (like you see in the McDonald's bathroom showing when the last time the bathroom was cleaned and by whom).

Figure 5.9 Cleaning tools.

There is also the concept of "cleaning as a form of inspection," which involves keeping the workplace clean, inspecting while cleaning equipment, and as a result, possibly finding minor problems during cleaning inspection. During the process, a greater emphasis is placed on the maintenance of machines and equipment.

The first three S's are the condition in which you want the workplace to be kept. In a way, they are like spring cleaning, except, if implemented properly, you do not have to do it every spring as the workplace is maintained at that level!

The next two phases, standardize and sustain, are all about *keeping* the workplace safe and organized.

Standardize

The fourth S is for standardize, which means creating a consistent way to carry out tasks and procedures—everyone does it the same (documented)

No.	5S Standardize: Job cycle chart Name: _____ Dept: _____ Date: _____										
No.	5S Job	Sort	Set	Shine	Standardize	Sustain	Continuously	Daily (AM)	Daily (PM)	Weekly	
1	Red tag (entire plant)	▓									
2	Red tag (cell/line)	▓					▓				
3	Inventory check		▓						▓	▓	
4	Tool check							▓	▓		
5	Wipe area						▓				
6	Vacuum area										
7	Machine clean inspection							▓	▓		
8	Degrease work area								▓		

Figure 5.10   Job cycle chart.

way. It is really a form of the standardized work that was discussed earlier in this chapter. The idea is to standardize how the first three S's are maintained.

For example, you may review the area weekly to see if anything needs to be red tagged and removed (sort out), check inventory levels for supplies at the end of a shift (set in order), and have a 5-minute cleanup at the end of the shift (shine). A good way to do this is to create a job cycle chart (Fig. 5.10), where duties can be assigned and communicated.

Many companies employ the use of a 5S board for an area (or in some cases, for a facility). The 5S board is an idea place to display things such as current 5S Audits, "before" and "after" pictures, housekeeping checklists, area maps, and job cycle charts.

Sustain

The final, and perhaps the hardest to accomplish, S is sustain. Sustain refers to making a habit of maintaining correct procedures over the long term. No matter how well we implement the first four S's, improvement gains may be lost and 5S doomed to fail without a *commitment* from everyone (management down to operators) to sustain it. 5S does not end at the conclusion of the 5S *kaizen* event. 5S must become part of a company's "culture" and become a habit to be successful.

There is no simple answer to be successful with 5S. It is a combination of communications, management support (including a Lean champion to spearhead the program and possibly a coordinator), culture, and rewards (everyone always wants to know "what's in it for me?").

Communications methods can include 5S Posters, "before" and "after" photo exhibits, 5S newsletters, 5S manuals (English *and* Spanish!), 5S events, competitions, 5S department tours, and success stories. I cannot tell you how many times I have heard the story, "We 5S'd our area and it worked fine for awhile, but the second shift messed it up."

5S should become an everyday activity, with daily cleanup and weekly 5S activities; it should become part of everyone's job descriptions, and be measured and displayed on the 5S board.

While this chapter covered some basic tools for Lean, especially 5S, which is a great place to start in an office or warehouse, the next chapter will get into some more advanced topics, which can have an even greater impact on the bottom line for your company.

CHAPTER 6

Advanced Lean Tools: It's Not Rocket Science

The nice thing about Lean as a form of continuous improvement is that all of the concepts and tools (even the more advanced ones covered in this chapter) are fairly easy to understand—it's not rocket science.

The merger of Lean and Six Sigma, which is a more quantitative tool to eliminate variability in a specific process, was mentioned in Chap. 2. But it is important to understand that the tools we covered in the last chapter and even the more advanced ones we will discuss here, are fairly easy to grasp (many of which are used in Six Sigma). It is actually the ability for people to change and commit that is perhaps the hardest part, and we will talk more about that in the "Keys to Success" in Chap. 10.

Batch Size Reduction and Quick Changeover

There are two critical concepts which go hand in hand in any Lean program. They are the ideas of *batch size reduction* and *quick changeover* (sometimes also referred to as *setup reduction*).

If you recall the difference between push versus pull production, you can better understand this topic. In push, you produce in large quantities to spread your fixed costs among a large number of items, thus minimizing your costs per unit. In pull, you schedule closer to what the customer actually wants (i.e., make what you sell). The ultimate goal is one-piece flow. While this may be unattainable, it is the direction that you want, and need, to head toward.

There are many benefits to this JIT approach. In Fig. 6.1, which compares the two approaches, you will notice a few things. First, you can see that smaller batches reduce the overall cycle time for any one item. In the

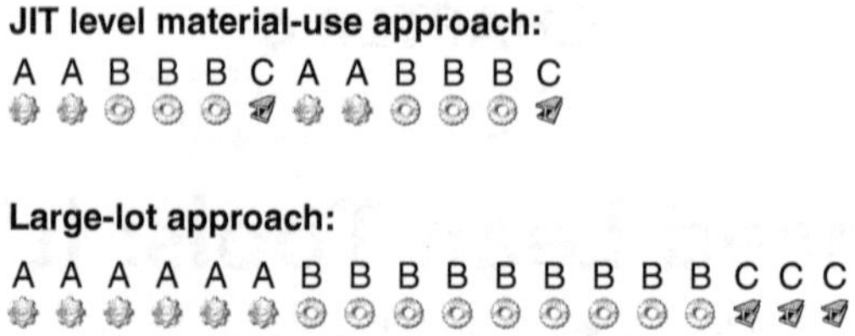

JIT level material-use approach:

A A B B B C A A B B B C

Large-lot approach:

A A A A A A B B B B B B B B B C C C

Figure 6.1 Push versus pull.

push approach, you have to wait for the large batches of other products scheduled before you even get to the one you are waiting for (item A in this example). You can also see that WIP is significantly reduced by the small lot approach. Finally, in the event that there is a quality problem that might affect the entire batch, it becomes less of a problem because of the batch size reduction.

Batch Size Reduction

The benefits of batch size reduction can include reduced lead times, lower inventory levels, more flexibility to meet fluctuating demand, better quality with reduced scrap and rework, less floor space used in production and storage, and thus lower overall costs.

In supply chain and logistics, there are the obvious results of batching in production to cover manufacturing wastes resulting in excess inventory, and in purchasing to obtain economies of scale (i.e., to get better pricing, which will be discussed more in Chap. 7), but it can also be seen in the office where batching typically occurs in the form of paper which can pile up in people's inboxes. There is a natural tendency to batch in an office as there usually some kind of setup for each type of activity, such as order processing, where files, forms, faxes, and reference materials are gathered before going to a specific computer screen. Each step in the process is typically done in batches and therefore ends up sitting in someone's inbox until they can get to it.

Quick Changeover

The primary obstacle to reduced batch size is changeover time and costs (note: the typical definition of a changeover in manufacturing is "the time from last good part to first good part"). The goal is to minimize changeover time and cost, so that smaller batches are run more frequently, resulting in better flow.

To go from one activity to another, whether on the shop floor, office or distribution center, requires some kind of changeover, which includes some kind of setup. As a result, batching seems like the most efficient way to do things (i.e., large batches equal fewer changeovers). Therefore, high setup costs encourage large lot sizes, and reducing setup costs reduces lot size and reduces average inventory.

If we can reduce the time and cost to changeover, batch size reduction can be realized. There is a concept called "single minute exchange of dies (SMED) which, while it literally applies to production operations involving a die, is used generally to refer to quick changeovers (or setup reduction), which result in smaller lot sizes and improved flow. Changeovers can vary in time from minutes to hours in manufacturing, but the idea is, through team-based continuous improvement, to keep reducing changeover time and cost so that things are produced in smaller batches or lot sizes.

Think of the race car pit crew and all that they do in a very short amount of time (Fig. 6.2). While the car is being serviced, it is not in the race, so the quicker the crew can get it back on the track, the better chance their driver has of winning the race. The same thing applies to changeovers and setups in business. If a piece of equipment is down, you are not able to make the product. The longer and more costly the changeover is, the less you want to do it, with the end result of large lot size batches. So if you can reduce the time and expense of setups, you can then changeover more often, making smaller batches and getting closer to one-piece flow.

Figure 6.2 Car pit crew.

Typical changeover tasks can involve preparation and adjustments, removing and mounting, measurements, settings and calibration, and trial runs and adjustments. If, as a team, you focus on waste in the current changeover process, it is not that hard to reduce the time it takes to changeover. Even a few minutes may be critical.

Most people involved in setups and changeovers feel that they are doing it the best possible way. However, the benefit of doing it in a *kaizen* event, with a mix of people from various functions, can shed new light on the subject. Typically, the team will observe and document the process (more than once if necessary) from beginning to end.

A useful tool in quick changeover is a setup analysis chart (Fig. 6.3). While a changeover is being observed (or videotaped as it may take hours), every step in the process should be documented including how long it takes, whether the step is "internal" (preparation while equipment is down) or "external " (while running) and distance traveled.

Some general keys to improving a setup are:

▲ Try to separate preparation or external setup from actual or internal setup and move as much as possible to external so you can shorten the changeover time.

▲ Move material and tools needed for the changeover closer to the actual spot it is needed.

▲ Standardize the actual process (and combine steps where possible) and tools used.

▲ Train operators and mechanics on procedures.

The net result should be an improved, shortened changeover.

In supply chain and logistics, as mentioned before, there is a large amount of batching of paperwork in an office, which if reduced can encourage improved flow and getting orders out faster, resulting in a shorter order-to-cash cycle.

In warehouse operations, there are setups everywhere, including receiving, picking, staging, loading, and shipping (especially for shift start-ups). Usually the first half-hour or more is fairly unproductive, so the more that can be standardized and visualized, the more productive personnel can be right from the start (especially temporary labor).

Figure 6.4 shows a receiving door that has been organized and visualized so that the operator can start up quickly as everything the operator needs to do the job is in place and ready to start unloading.

Setup Analysis Chart

Seq. #	Element description	Current method					Proposed method				
		Internal	External	Duration	Distance traveled	Improving ideas	Internal	External	Duration	Distance traveled	Comments

Figure 6.3 Setup analysis chart.

Figure 6.4 Organized receiving door.

Kanbans

A key to successfully going from a push to a pull environment is the use of *kanbans*.

A pull system typically uses signals to request production and delivery from upstream stations (it might be a card with replenishment information, or as simple as a line on the wall; for a simple example, see Fig. 6.5). Upstream stations only produce or replenish when signaled.

Figure 6.5 Simple *kanban*.

The tool to execute this process is called a *kanban*, which is essentially a way to control the flow of materials and other resources by linking functions with visual controls. Only what has been consumed at the demand by the downstream customer is replaced. This determines the production and replenishment schedule.

By pulling material in small lots, inventory cushions are removed, exposing problems and emphasizing continual improvement

There are many benefits to a pull system, including reduced cycle time, less orders "dumped" downstream creating excess WIP, less reliance on a forecast, and short lead times for customized products or services.

The most common thing you may hear about *kanbans* that failed is that "we had one but it stopped working." One of the main reasons for this is the establishment and maintenance of reorder points and reorder quantities. Lead time and demand rate to replenish can determine "when" to order and the Economic Order Quantity (EOQ) model, which minimizes inventory holding and ordering costs, can be a way of determining "how much" to order. In many cases, an item's demand may have "peaks and valleys." As a result, it is necessary to continually review *kanban* order points and order sizes, as they may need to be adjusted based upon the current demand rate.

Many companies also try to use a "one size fits all" type of model when establishing *kanban* reorder points and quantities. This of course dooms the program to failure as each item has its own individual demand characteristics and needs to be looked at individually.

In supply chain and logistics, *kanbans* have many applications. Starting with the obvious replenishment of raw materials with vendors. This replenishment method can be initiated manually where a *kanban* card is *pulled* when inventory is running low. The *kanban* card has basic item information such as reorder quantity, pricing, etc. needed to place a reorder manually or electronically based upon a predetermined reorder point. The extreme of this is called *vendor-managed inventory* (VMI) (which will be discussed in more detail in Chap. 13), where the supplier takes full control over replenishing materials, saving the customer the costs of monitoring inventory and placing orders. Typically, the vendor either makes regular visits to check on inventory levels, or the information is transmitted electronically to the vendor (there are now even vending-type machines to dispense and automatically reorder parts and hardware directly from

Figure 6.6 Simple *kanban*.

the vendor). This results in orders being pulled in small lots, more frequently than was the case before the VMI (and also eliminates the need for you to check inventory and place purchase orders).

A *kanban* can also be used to replenish supplies and packaging materials. A good use for *kanbans* in distribution centers is to replenish supplies, such as labels, tape, and corrugated boxes in an office or work area. It can be as simple as a line on a wall to determine reorder point (see Fig. 6.6).

Kanbans help to enable the concept of *point-of-use storage* (POUS), which is the idea of having things you use more often closer to you (and typically at waist level) and the things you use less often, further away and higher up. The *kanban* can be used to keep minimal amounts of raw materials and supplies nearby, without taking up a lot of space in the work area where space is at a premium and efficiency is everything. It also simplifies physical inventory tracking, storage, and handling and is a foolproof way to ensure that an area never runs out of needed materials or supplies.

Quality at the Source

Another Lean concept is *quality at the source* (also known as *source control*). The idea here is that the next step in any process is the customer, and you want to make sure that you deliver perfect products to that customer.

A way to help ensure this is through the use of a *poka yoke,* a Japanese term for mistake-proofing. A poka yoke is a way to make so that it is virtually impossible to pass on a defective part or piece of information from one process to another (think of the saying "you can't put a square peg in a round hole").

This does not just pertain to physical product, where you can create a foolproof device to ensure that the correct product is passed on (e.g., creating a right-angle "jig" to test the product before passing it on). It can be used for information by limiting the choices on a form or screen, for example.

Quality at the source is typically used in conjunction with *a total quality management* (TQM) program. TQM is similar to Lean Enterprise in that it is a team-based program spanning the entire organization, from supplier to customer, and requires a commitment by management to have a long-term, companywide initiative toward quality in all aspects of products and services as defined by the customer.

There are seven tools of TQM. They are: continuous improvement, Six Sigma, employee empowerment, benchmarking, Just in Time (JIT), Taguchi concepts (specific statistical methods developed to improve quality), and knowledge of TQM tools such as Pareto charts and cause-and-effect or fishbone diagrams. Some of the *seven tools* of TQM are also found in Lean thinking (continuous improvement, empowerment, and JIT).

Work Cells

Another powerful tool that can significantly impact the efficiency of the workplace is a work cell. Work cells rearrange people and equipment that would typically be located in various departments into one group so that they can focus on making a single product or providing a single service or a group of related items or services. It is not a new idea, as it was originated in 1925 by R. E. Flanders.

Work cells are typically layed out in a horseshoe-type shape, which allows for a more efficient use of work space and equipment and is conducive to one-piece flow and, as a result, less WIP. There is also the benefit of needing less workers, as each worker in the cell is able to do all of the activities required to produce or assemble the product or to deliver the service. Employees in a work cell typically have higher morale as a result of greater participation in the entire process.

The typical first step is to identify families of products or services (a "family" should have most , but not necessarily all, of the same steps). As a result of the *job enlargement,* a typical feature in a work cell where employees have multiple responsiblities is that there is a great deal of training required, which results in a lot of flexibility in the cell. There is often an opportunity to use poka yokes to ensure good quality as well.

Balancing a Work Cell

It is then important to calculate the *takt time* for the product or service family (total work time available/units required) in order to balance the activities in the work cell so that materials or information can flow. So, for example, if the daily demand for a product (on an 8-hour shift assuming no breaks in this example) is 800 units, the takt time is 100 units/hour or 1 unit every 36 seconds. Each activity in the cell should be capable of making and passing 1 unit every 36 seconds in order to be balanced.

This information can help to identify *bottlenecks* in the process where one or more components or resources limit the capacity of an entire system, so as to make sure the cell is balanced properly (i.e., that each step in the cell is capable of processing one unit at the required rate of takt time). In the event that it is not balanced, having flexible, cross-trained employees can help address bottlenecks; in the case of machine bottlenecks, other approaches may be necessary, such as running overtime, speeding up, or adding equipment.

A time observation form (Fig. 6.7) is a useful tool when looking to staff and balance any multiple-step process. You take multiple measurements of the time to do each activity to see if the entire process is balanced and if there are any bottlenecks (versus the takt time). It can also help to see if you have the proper number of people involved in the process (total observed time to complete all activities for one part divided by takt time).

Work cells can be appropriate not only on the shop floor, but in the office and warehouse. For example, in the office for a distributor, there are various functions that are required in order processing, including receive order, check credit, review and enter order, reconcile and confirm order, and finalize and release order. Typically these activities are done by

Time Observation Form
Use this form to help design a specific workstation within a cell

Step 1: Down the left side of the form, list the component tasks in the order in which they are performed.
Step 2: Observe each component task multiple times.
Step 3: Average the times for each component and the total cycle times.
Step 4: Estimate the value-added and non-value added time for each component task.

Time Observation Form												Date:		Event or observer:	
												Model:		Takt time:	
#	Component task	1	2	3	4	5	6	7	8	9	10	Avg. time	Value added time	Non-value added time	Comments and observations

Figure 6.7 Time observation form.

different people, who are possibly in different departments. This activity can take a day or longer, while the actual value-added activities may only take 30 minutes or so. Along the way, there may be batching of orders, waiting for approvals, and a lot of walking around.

If this function were to be set up in one work cell performing all of the activities, the entire process could now take less than 30 minutes for each order using fewer employees (who have had their jobs expanded). The end result is that the orders are released for picking and shipment much quicker, thus speeding up the order-to-cash cycle for the business.

If you are implementing a work cell, you should realize that there may be higher wages involved as a result of more responsibilities as well as more training, but it can be well worth it overall.

In the distribution center itself, there may be more limited oppor-tunities, but they are typically found in areas like packaging or value-added activities performed by 3PLs such as packaging of kits for a customer.

Total Productive Maintenance

The final major concept in Lean that I would like to cover is that of *total productive maintenance* (TPM), which focuses on equipment-related waste. Equipment maintenance (or lack thereof) is often an overlooked area of waste. In fact, studies have shown that most manufacturers (70 percent or so) operate under what is commonly called *breakdown maintenance*. You would not treat your car like that. Every 3,000 miles or so, you take your car in to change the oil and air filter, lubricate vari-ous parts, check fluid levels, etc. This is called *preventative maintenance* (PM). You do not wait until the transmission drops out on the road before checking the fluid, gears, etc.

There is another type of maintenance called *predictive maintenance,* but that is typically used after a good PM program is in place. In predictive maintenance, tools are used to check temperature, vibrations, etc. to see if some corrective action is necessary.

TPM, in and of itself, is not a PM program. It can, however, result in putting such a program in place.

Overall Equipment Effectiveness

Basically, a piece of equipment is observed (and possibly videotaped) dur-ing an entire shift to come up with an *overall equipment effectiveness* or *OEE* percentage (Fig. 6.8). According the various studies, typical compa-nies average in the 70 percent area. That means that there is room in most companies to reduce or eliminate equipment-related waste to increase throughput and quality.

During the shift, observations are made as to how the equip-ment is running. The observations are broken into three categories: performance efficiency, availability, and quality. Within each cate-gory are specific reasons for slowdowns, stoppages, breakdowns, and

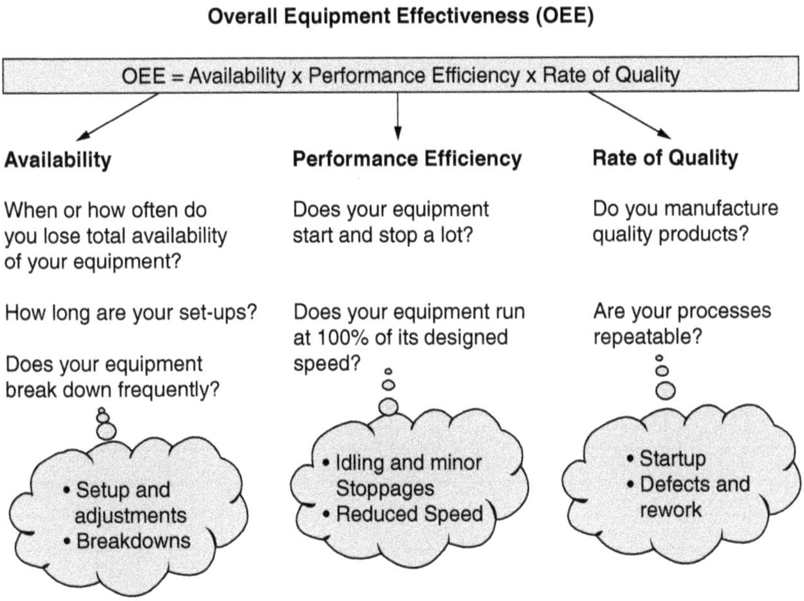

Overall Equipment Effectiveness (OEE)

OEE = Availability x Performance Efficiency x Rate of Quality

Availability	Performance Efficiency	Rate of Quality
When or how often do you lose total availability of your equipment?	Does your equipment start and stop a lot?	Do you manufacture quality products?
How long are your set-ups? Does your equipment break down frequently?	Does your equipment run at 100% of its designed speed?	Are your processes repeatable?
• Setup and adjustments • Breakdowns	• Idling and minor Stoppages • Reduced Speed	• Startup • Defects and rework

Figure 6.8 OEE calculation.

quality issues. This information is put on an OEE observation form (Fig. 6.9).

After the shift is over, the various observations are tabulated and separated into each of the three major categories to come up with an overall OEE percentage for the piece of equipment observed. As with a changeover *kaizen* event, a video is helpful in TPM to see more detail.

An *analysis of the major losses* (Fig. 6.10) is then performed, which shows the most productive place to begin an improvement process. In this example, you can see that most of the major losses are breakdowns, thus narrowing down where you should look to improve equipment peformance.

Once you have determined the major source of waste or losses, you can use a Pareto analyis to get to the root cause of the waste (Fig. 6.11). In this example, over half of the breakdowns are in the load arm. By solving this one issue, you can make major progress.

The team, which should be made up of a variety of people, including maintenance personnel, engineers, and leads, should meet to discuss the major causes of waste and how to reduce or eliminate them. The

OEE Observation Form

Date:_____   Observer:_____

Equipment #_____   Description:_____   Dept:_____

Start from	End to	Performance efficiency						Availability				Quality	
		Idling and minor stoppages						Breakdowns					
To	Running time	Chips	Jam	Insert	Other	Reduced speed	Lube		Other	Setup and Adj	Startup	Defects and rework	Comments

Figure 6.9 OEE observation form.

solutions can range from simple, inexpensive solutions called *countermeasures* (e.g., putting a filter over a motor to reduce the buildup of dirt and grease coming in to it), to daily, weekly, and monthly PMs (some performed by the operator and others by mechanics).

All of these solutions should be put into an improvement plan, which describes the problem you are going to improve, a list of questions you have, a summary of action steps, and a plan to monitor improvements.

Analysis of Major Losses

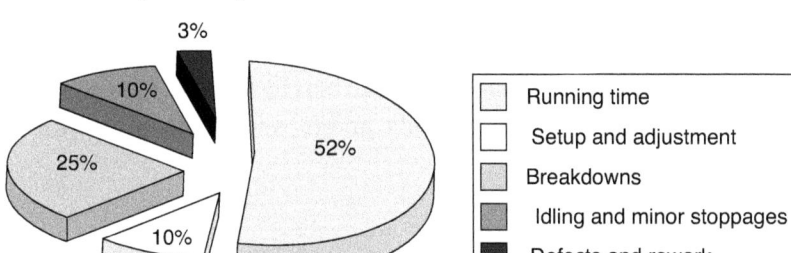

Figure 6.10 Analysis of major losses.

TPM can be applied in the office as there is plenty of technology that can affect productivity, such as computers, copiers, and fax machines, to name a few. Same goes in transportation and distribution where everything from carousels, forklift trucks, automated storage and retrieval systems (ASRS), pallet wrappers, and radio frequency (RF) devices can affect productivity.

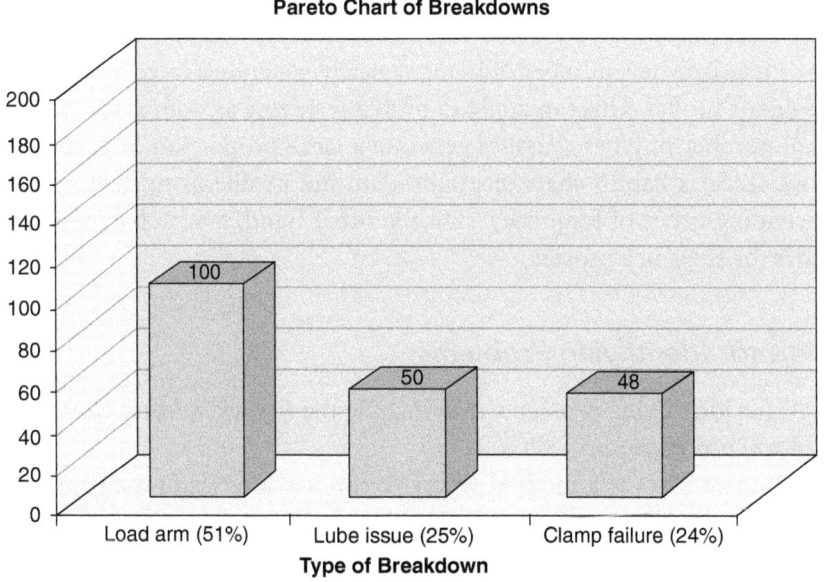

Figure 6.11 Pareto chart of breakdowns.

Lean Analytical Tools

There are many analytical tools that can be used in Lean, some of which are used in TPM as mentioned earlier. They can best be divided into categories of tools for (1) gathering, (2) organizing, and (3) identifying problems.

Tools for Gathering

The tools for gathering data include simple *check sheets, scatter diagrams,* and *cause-and-effect (or fishbone) diagrams.*

Check sheets are simple "chicken scratches" to organize data; scatter charts are basically a graphical view of the relationship between two variables; and fishbone diagrams (called that because they are shaped like a fishbone) show process elements or causes of an outcome or effect. These are useful tools for teams to brainstorm improvement ideas by first trying to come up with possible reasons for waste.

Tools for Organizing

The tools used for organizing data include Pareto charts (Fig. 6.11) and flowcharts.

Pareto charts use the 80/20 rule (discussed previously), which states that a relatively few number of items typically generate a large percentage of sales or profits. This can apply to problem solving as well, as typically a small number of types of issues generate a large proportion of problems or waste. So a Pareto chart or graph identifies problems or defects in a descending order of frequency. On the other hand, a flowchart visually shows the steps in a process.

Tools for Identifying Problems

Tools for identifying problems may include the *five whys, histograms,* and *statistical process control* (SPC).

The five whys is a method whereby you ask a series of questions in order to get to the root cause of a problem or defect. The idea is that the answer to each question leads you to ask another "why" until you get to the source of the problem.

Histograms are a graphical way to show the distribution of the frequency of occurances of variables, such as problems or defects.

SPC is a huge subject, of course, and one of the major tools used in Six Sigma to minimize variability in a process. In general, SPC utilizes both statistics and control charts (that indicate whether a process is in a state of statistical control) to tell when you need to take corrective action. It can be used for process improvement as well.

There are four major steps to SPC. First, if you don't measure a process, you can't control and improve it. A *control chart has upper* and *lower control limits* (UCL and LCL). If samples are within those bounds, then the process is viewed to be "in control." Second, before making a change to a process, make sure that you have an *assignable cause*. Third, you should try to reduce or eliminate the cause of the problem or defect and fourth, restart the revised, improved process.

In terms of supply chain and logistics, many of the Lean Six Sigma tools mentioned above can be used to identify, analyze, and minimize variation in areas such as inventory control, forecast, customer demand volatility, and on-time delivery, to name a few.

For example, a client of mine manufactures helicopter accessories. The team identified waste in their receiving area that slowed down the receiving and put-away process. They determined that the wastes were caused by three major categories: suppliers, resources, and workers. To both validate this and gather data, the team had receiving personnel collect data for a month. They then used the Pareto principle to segment the major wastes in each of these categories, and then a kind of informal "fish bone" to determine the major causes of the largest waste within each category (see Fig. 6.12). So in the example shown, the data indicated that the largest supplier-caused waste was missing paperwork. Then, they drilled into the collected data to determine the major causes of missing paperwork were missing certification (58 percent of the occurances) and blue/yellow tag (20 percent). Subsequent next steps and action items were taken to work with suppliers and internally to eliminate the two largest causes of missing paperwork, which made up 78 percent of the identified waste—a real improvement.

In general, using many of the tools described in this chapter, Lean Six Sigma teams (and teams in general) can follow five problem-solving steps to quickly identify root problem causes, develop solutions, and put in place

ABC Company
Receiving Kaizen Event–Supplier

% of Waste	Description
80%	Missing paperwork
15%	Incorrect/Incomplete paperwork
5%	Early delivery

% of Cause for Waste	Types of Missing Paperwork	Comments
58%	Certification	
20%	Blue/Yellow Tag	
15%	Age control	Biggest waste of time, but fewer occurances
1%	Check sheet	
1%	Packing slip	

Summary of Next Steps:

(1) Receiving (John Smith and Joe Jones) will track occurances of each type of missing paperwork for 30 days (including type of missing paperwork, supplier, etc).

(2) John and Joe will analyze findings and take corrective action for major causes of waste.

Summary of Findings to Date:

(1) Review of one month's tracking found that main causes of missing paperwork were:

- Track card not signed/missing
- Ship to WIP not transacted
- Wrong Rec case (buyer not updating)

Also:

- Materials not issued
- Overshipped

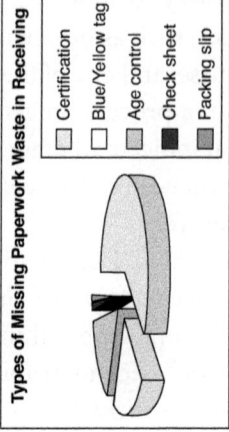

Receiving Wastes–Supplier

- ☐ Missing paperwork
- ☐ Incorrect/Incomplete paperwork
- ☐ Early delivery

Types of Missing Paperwork Waste in Receiving

- ☐ Certification
- ☐ Blue/Yellow tag
- ☐ Age control
- ■ Check sheet
- ☐ Packing slip

Figure 6.12 Receiving *kaizen* event example.

procedures that maintain those solutions. A formal Six Sigma model used for process improvement is known as DMAIC, which stands for:

- **Define**—Identify the customer requirements, clarify the problem, and set goals.
- **Measure**—Select what needs to be measured, identify information sources, and gather data.
- **Analyze**—Develop hypotheses and identify the key variables and root causes.
- **Improve**—Generate solutions and put them into action, either by modifying existing processes or by developing new ones. Quantify costs and benefits.
- **Control**—Develop monitoring processes for continued high-quality performance.

All of the tools and methodologies discussed in this chapter are very useful for getting to the root cause of problems and defects, but in order to make them work for you, there has to be a team-based continuous improvement *culture* in place, which is described in Chap. 10.

CHAPTER 7

JIT in Supply Chain and Logistics: This JIT Is Good

Much has been made of the usefulness of Just in-Time (JIT) as a Lean manufacturing tool through the years. It has been applied throughout the manufacturing process to minimize waste with much success. In terms of supply chain and logistics management, however, there is still tremendous untapped potential. As result, it is a good idea to devote an entire chapter to the subject.

As we know, inventory acts as a buffer between your internal processes—between you and your customers and between you and your suppliers. It is kind of a "necessary evil" because of variability and uncertainty in the system. In terms of your supply chain management (SCM) function, it is in place to compensate for forecast error, variability (and length) of lead times, inventory inaccuracy, order size minimums, damages, pilferage, shipment errors, etc.

While there has always been an emphasis on cutting SCM costs as it is easier to reduce costs a little than increase sales a lot for the same contribution to the bottom line, it has gotten harder and harder to "get blood from a stone." That's where Lean and continuous improvement can help your organization to keep its eye on the prize.

Areas of Focus

With today's global sourcing, there is an even greater need to efficiently manage inventory in the supply chain. As a result, there is a tendency to carry excess safety stock to help compensate for this. This excess safety stock actually costs the company money via carrying costs, which can be in excess of 20 to 30 percent of the cost of the inventory, so it is important to get to the root cause of the variability and reduce or eliminate it so as to minimize this inventory.

Typically, the sourcing or purchasing manager tends to try to reduce the company's material costs as much as possible. This leads to buying in larger volumes. The same goes for the logistics manager, who also focuses on low cost and reliability in the transportation and distribution network.

However, when your operations are JIT-based, you still need to focus on cost and reliability, but other dimensions are now added, such as flexibility and agility to meet rapidly changing customer shorter-term demand.

In a JIT environment, the focus is on lower *total* cost, not just lower per-pound or unit raw material cost. You need to look at other aspects, such as transportation, handling, storage costs, etc., and at new types of relationships with suppliers (including material, transportation, distribution, and 3PL partners). Logistics managers must consider shipping in smaller and smaller quantities, more frequently using less-than-truckload (LTL), 3PLs, freight forwarders, and consolidators. By focusing on lower total cost, you can somewhat offset the higher transportation and handling costs of smaller loads and get the benefits of lower storage and overall carrying costs.

You now need to develop more of a partnership with your suppliers where they can do more than give you the lowest price and best quality. Again, you need to consider total cost based upon not just unit price and transportation costs, but overall carrying costs and the price of inflexibility to your company. You should be able to set up JIT relationships in which materials can be resupplied based upon downstream "pulled" requirements.

There is, of course, a lot of risk with the extreme partnering of single sourcing" material as we saw during the 2011 earthquake and tsunami disaster in Japan. Many manufacturers and distributors are being affected by shortages as a result of this horrific event as first-, second-, and even third-tier suppliers have parts that are sourced in Japan, especially in the area of high-end highly technical parts like semiconductors, which also are very small and light. Additionally, they typically ship by air so manufacturers don't have to keep much of a buffer for those products.

The lesson to be learned here is that JIT still makes sense, but it has to be applied strategically. If, for example, you purchase some kind of semiconductor (20 percent of which are made in Japan), you should consider having a secondary supplier outside of Japan, even though you may pay a bit more per unit.

Network Design

If you have a fairly complex supply and demand network, it is usually best to first look to optimize your overall distribution network design based upon these new JIT goals to make sure that your entire network (i.e., procurement, manufacturing, transportation and distribution) is designed to support this strategy (note: it is not a bad idea to do this on an annual basis, as it can lead to significant cost reductions and improved service levels).

Typically, a network analysis can be a complex undertaking done by consultants. However, with the advent of improved, lower-cost technology, the cost of doing this has dropped dramatically. Some examples of this software include IBM's ILOG, Axxom's ORion-PI, Oracle's Strategic Network Optimization, MicroAnalytics Opti-Site, to name a few (visit http://www.lionhrtpub.com/orms/RD/products.html for a larger list).

Integration of Resources

In *Principles of Supply Chain Management—A Balanced Approach*, Joel Wisner et al. point out that there is really an "evolution" that a firm must go through to institute Lean with JIT internally first, then vendor-managed inventory (with suppliers) and *quick response* (QR)/*efficient consumer response* (ECR; a form of VMI used to manage customer's inventory of your product), which are really external expansions on the idea of JIT. The stages of JIT are:

- ▲ Stage 1—Firm is internally focused and functions are managed separately. This functionally focused silo effect is reactive and short-term goal oriented.
- ▲ Stage 2—Firm integrates efforts and resources among internal functions.
- ▲ Stage 3—Firm links suppliers/customers with firm's processes.
- ▲ Stage 4—Firm broadens supply chain influence beyond immediate or first-tier suppliers and customers. [Wisner et al., 2009]

There are significant challenges in getting from the first stage to the fourth. We have to realize that managing a supply chain is not easy, as it is usually a geographically dispersed network with conflicting objectives across the supply chain. There are many conflicting objectives among the participants (e.g., manufacturers produce and ship in relatively large quantities and retailer wants small shipments fairly frequently) with a lot of uncertainty and risk inherent in the system. Also, information tends to get distorted as it is passed among the supply chain partners (think of the "telephone" game you played as a child).

Lean supply chain relationships develop where suppliers and customers work to remove waste, reduce cost, and improve quality and customer service. JIT purchasing includes delivering smaller quantities, at right time, delivered to the right location, in the right quantities. Firms develop Lean supply chain relationships with key customers. Mutual dependency and benefits occur among these partners.

Walmart and Dell: Examples of JIT in the Supply Chain

Walmart and Dell are two prime examples of a Lean supply chain. It is best to look at how they accomplished this so that we can all learn where to look in our own businesses.

Walmart was one of the first retailers to collaborate with suppliers in the late 1980s/early 1990s with their *Retail Link* system. Having worked for several of Walmart's larger suppliers, Unilever and Church and Dwight (Arm & Hammer), during that period, I can discuss my experience with Retail Link firsthand.

Walmart's Retail Link was an early way of sharing *point of sale* (POS) data, distribution center inventory levels and shipments to retail information, as well as working collaboratively to improve forecasts. In addition, Walmart is a heavy user of *cross docking*, which involves the unloading of materials from an inbound truck or railroad car and then loading these materials directly into outbound trucks, trailers, or rail cars, with little or no storage in between (typically, this all happens within 24 hours). They also use *radio frequency identification* (RFID), which uses a radio frequency (RF) device to read data from an electronic tag attached to an object, for the purpose of identification and tracking.

More recently, according to the article "Walmart takes back its Supply Chain—IT in the Spotlight" by Frank Hayes, Walmart is planning on increasing the use of its own vehicles to pick up merchandise right at the suppliers' shipping docks. While this is not new (it is known as "backhaul"), it can help them to cut their wholesale costs by as much as 6 percent and, perhaps more importantly, get better control over their inbound inventory (i.e., when and how it arrives, and how quickly it can be turned around). Currently, Walmart tracks pallets as soon as they are delivered to the distribution center (DC), where the RFID tags are scanned and the information linked to EDI information, which shows what should have been shipped. In the case of backhauls, with Walmart picking up the shipment, the RFID tag will be scanned at the manufacturer's dock and any errors will be caught right away, thereby implementing "quality at the source" and a kind of poka yoke into the process. This will give Walmart better inventory accuracy, visibility, and predictability. [Hayes, 2010]

Dell is a great manufacturing example of a Lean supply chain in the age of mass customization. They can take a customer's order and assemble and ship it within 24 hours and get inventory turns as high as 90+ times/year (has varied from 40 to 140 turns/year).

Dell has perfected "strategies such as direct-to-consumer sales, bare-bones inventory, reverse cash conversion (which pays suppliers after, rather than before, receiving payment from customers), and a requirement that suppliers retain possession of parts until the last possible minute have enabled Dell to build a super-lean business model that has upended high-tech manufacturing much in the same way Wal-Mart changed retail." [Risen, 2006]

Dell has a very tight relationship with its vendors. They develop very strong relationships with a few key suppliers and have "virtually integrated" their suppliers so it appears as though the vendors are an extension of the Dell Corporation. Parts are provided JIT to the point that the exact number of items needed are delivered daily, and in some cases even hourly.

In fact, for its various worldwide manufacturing facilities,

Dell does not buy raw materials and components and maintain inventory. Dell's vendors use third party service providers to set up logistics parks and distribution warehouses close to Dell's plants and deliver materials just in time to the plant against an order for production, which is triggered based on an order confirmed by the customer on the internet.

Dell has appointed freight forwarders such as DHL, CEVA, Panalpina, UPS etc...to pick up shipments from vendor locations, transport the collected shipments by road and consolidate inventories of all vendors in the freight forwarders consolidation warehouses situated at the gate ways in each country and ship out cargo by ships to the port of destination or airfreight shipments to the plant locations after completing exports and customs clearance formalities on behalf of vendors.

While the shipments are in transit, the freight forwarders electronically transfer shipment information and documentations to their overseas offices or agents at the destination and keep Dell and vendors informed of the status of shipments.

Freight forwarders at the destination ports file advance shipment documents with customs and on arrival of cargo, complete customs formalities and custom cleared cargo is then transported to freight forwarders warehouse or customs bonded warehouse or to another designated third party warehouse which houses all inventories meant for Dell.

The third party service provider who manages the inventories in his warehouse receives the cargo, unpacks the shipments from bulk skids to individual carton level and completes inbound formalities including updating of inventories in its system and stocks the materials in designated rack locations. Both vendors and Dell are continuously kept informed of the data regarding shipments and stocks. The warehouse stocks inventories in the name of various vendors at SKU level. Most of the times, these warehouses are situated adjacent to the plant or at close proximity. Upon receiving a production order from Dell, as per Bill of Material received through DELL ERP system, items are picked up, loaded into the supply cages and trays as per pre determined design and delivered to the plant after completing documentation and system entries to remove inventory from its system held in vendors name, invoice raised and physical delivery accompanied with documents completes the supply chain cycle of Raw material supply.

The revenue recognition happens when material is transferred out of the warehouse and its system and invoiced to Dell. [www .managementstudyguide.com, 2011]

This gives a clear example of how Dell manages to maintain visibility and reliability in its supply chain and logistics network, outsourcing activities that are beyond the boundaries of their core competencies.

Visibility and Reliability

Visibility helps supply chain members to see and manage the flow of products, services, and information, in near-real time, from customer to supplier. It is a seamless integration in which access to information on inventory, product availability, and order status enables the supply chain to execute as if they were a single "virtual" entity. Visibility allows supply chain managers to see the flow of materials and orders and to better manage capacity and resources, resulting in better reliability of the system.

According to Michael Dell of Dell Computer, "The greatest challenge in working with suppliers is getting them in sync with the fast pace we have to maintain. The key to making it work is information." If we go back to the SCOR model that we discussed earlier in the book, recall that there is a broadened scope of information which must be communicated and coordinated. In order to control costs, risks, and time, we need added visibility to understand the various relationships, constraints, and bottlenecks. To have access to that information is to have visibility. [Magretta, 1998]

As Gary Carleton points out in "Wringing Cost out of the Supply Chain," "a successful JIT inventory strategy isn't something that can be done alone," especially in the case of small- and medium-sized enterprises (SMEs). He points out that there are "black holes" in the global supply chain where you can lose sight of your inventory. So it is important to partner with logistics providers that have tools and technology to give you better visibility enabling your company to become more flexible to meet changing demand and other nonplanned events. [Carleton, 2011]

If you always look for the cheapest transportation provider, you may not get the reliability and visibility that more expensive ones may be able to provide with technology and communication.

In "Supply Chain Velocity: Shifting into Overdrive", Joseph O'Reilly points out how important it is for many companies to track and trace shipments in transit, especially for customers who demand timely service. He mentions how a logistics provider, Pegasus Logistics Group, finds that "pressures often come from dynamic changes in customer demands as they try to capitalize on getting products to market" and that "upstream and downstream visibility into our clients' orders provides the ideal awareness of due dates and delivery schedules", so they "…can match this information with carrier capabilities to execute timely deliveries. Managing the suppliers' order release process is one example where we can improve efficiency."[O'Reilly, 2010]

Efforts to share information have been implemented through the years, such as EDI, but the reality is full visibility is elusive, as manufacturers, their suppliers, customers, and third-party partners each have implemented various "best of breed" and/or ERP systems. So "true" visibility, especially in real-time, is not a reality yet. Manufacturers each have their own systems (some still "homegrown" and primarily paper-based). Many manufacturers have spent millions on integrated systems and are disappointed, but they do not want to give up on them as a result of the large investment. Sometimes it is the software's shortcomings or a culture that is resistant to change.

The reality is, though, that it is really the data, not the software, and that is where Internet-based shared networks can help (e.g., software as a service or SaaS). This may be the future of technology and the ultimate way to increase visibility into the complex supply chain and root out waste.

Most manufacturers are lucky if they have 50 percent visibility into their supply and demand chain. As a result, it is harder to execute JIT systems, and there tends to be a buildup of "just in case" buffer inventory. With better visibility, where we can trust the data, we have a better handle on what is in stock and in transit. [Krizner, 2010]

We will talk more about how technology enables a Lean supply chain later in Chaps. 12 and 13.

Cross Docking

Cross docking, discussed briefly earlier in this chapter, is a logistics activity that attempts to reduce costs and total lead time and better synchronize with actual demand and therefore is a great tool to enable a JIT process. When successful, it can enable a JIT supply chain strategy in a fairly "extreme" form. It does this by breaking down received items on the loading dock and immediately matching them with outgoing shipment requirements, as opposed to stocking the items in warehouse locations and returning to pick them for orders at a later time.

Cross docking essentially eliminates (or minimizes) the inventory holding function of a warehouse while still allowing it to serve its consolidation and shipping functions. The idea is to transfer incoming shipments directly with outgoing trailers without storing them in between and, as a result, is very dependent on transportation (inbound and outbound).

Typically, the goods arriving from a vendor already have a customer or retail destination assigned. In predistribution cross docking, the customer is assigned before the shipment leaves the vendor. In postdistribution cross docking, material is allocated for the stores at the cross dock itself.

There are many advantages to cross docking. They include:

▲ The minimization of warehousing and economies of scale in outbound flows (from the distribution center to the customers).
▲ The costly inventory function of a DC becomes minimal, while still maintaining the value-added functions of consolidation and shipping.
▲ Inbound flows (from suppliers) are directly transferred to outbound flows (to customers) with little, if any, warehousing.
▲ Shipments typically spend less than 24 hours in the distribution center, sometimes as little as 1 hour.

Cross docking can be a challenge to execute and has been used more in the retail sector but also has applications in manufacturing and distribution. To execute the cross docking concept most effectively, you should have automated material handling including a conveyor system with bar code scanners (even better if you have RFID from the supplier), a *warehouse management system* (WMS) to help control the flow of material and information internally, good quality control to avoid delays, bottlenecks, or the costs associated with shipping inferior product, and above all, a good working relationship with your supply chain partners.

While many companies have implemented JIT in their manufacturing function, there is still plenty of room for it to be implemented "outside the four walls" in areas such as procurement, transportation, and distribution, in order to have a leaner supply chain.

CHAPTER 8

Lean Warehouse: Low-Hanging Fruit

Lean is still in its early stages in supply chain and logistics, so it is sometimes difficult finding a place to start it. A place that many companies have found as a good place to start is the warehouse, which was discussed briefly earlier in the book.

In the 21st century, the warehouse is becoming a strategic tool to be used for a competitive advantage. Warehouses today are distribution centers supporting a JIT supply chain that is low cost, flexible, and efficient, especially in the rapidly growing world of e-commerce. E-commerce growth affects both the warehouse and the inbound and outbound logistics that support the facility.

Lean Thinking in the Warehouse

In "The Skinny on Lean," Peter Bradley cites the following five-step process as a guide to implementing Lean principles, which can be applied to the distribution environment as well as manufacturing:

▲ Identify what your customers expect and determine what value you add to the process. For distribution and logistics, that usually means greater velocity. What it doesn't mean is a lot of handling. Distribution people assume all the handling they do adds value, but customers don't see it that way. "No customer asks if a product has been touched a lot," Womack says. "Most people just want their product. All those touches from a customer standpoint are irrelevant. From an end customer standpoint, less logistics is better."

▲ Plot the value stream. Identify all the steps involved in moving goods through the system. Womack and Jones encourage the use of value-stream mapping—literally diagramming all the steps in the distribution process, from order to delivery. That diagram may help you spot activities that add no value so that you can eliminate them.

▲ Make the process flow. Dismantle any roadblocks that prevent the free flow of materials through the facility.

▲ Pull from the customer. The lean system is a pull system, drawing materials and merchandise into the distribution network based on what customers want (not on hazy forecasts).

▲ Pursue perfection. Root out any remaining waste. Then do it again, and again, and again."[Bradley, 2006]

The fact is that most Lean concepts can work well in the warehouse, especially 5S (usually the first activity to do as a good foundation for a Lean program), value stream mapping (VSM), team building, *kaizen*, problem solving and error proofing, *kanbans*/pull systems, line balancing, and cellular applications and general waste reduction.

At first glance, many warehouses are very neat and organized, at least in the case of "pure" distribution companies, and perhaps less so at manufacturers (not their area of expertise!). However, once you "get under the hood," there are plenty of opportunities to be found.

"Assembling" Orders

Warehouse operations seem to be very active, with people and equipment in constant motion. However, this does not mean they are productive. Orders do not necessarily keep moving. They tend to pile up and sit, waiting between processing steps, causing clutter and taking up space (all forms of waste!). An analysis done around 2006 of a distribution operation, as described in *Are Your Warehouse Operations Lean?* by Ken Gaunt showed that a typical order was only being worked on 38 percent of its cycle time; 56 percent of the time orders were idle while the remaining 6 percent involved employees dealing with problems such as waiting for equipment, computer issues, interruptions, and blocked aisles.

To radically improve this type of performance, warehouse orders should be looked at as being "assembled" in the most efficient manner, minimizing non-value activities including delays in receiving, putting away, picking, packing, and loading, as well poor picking paths, wasted motion, congestion, and poor equipment condition and availability.

For example, the orders could be assigned based on the amount of time or "batches" that it takes to pick line items (established in time motion studies), instead of just giving an entire order to a picker. Pickers would be assigned zones. They would then feed workstations in regular intervals to keep products flowing and so that packers and loaders are not kept waiting.

In general, you want to look at aisle and rack layout to improve space utilization, make sure products are arranged so that the most frequently used items are closest to shipping to reduce travel distance; heavily use visual systems for aisles, racks, products, and workflow; avoid cluttered and blocked aisles; and create housekeeping systems to improve efficiency by ensuring tools and equipment are available when and where they're needed. [Gaunt, 2006]

Value Stream Mapping in the Warehouse

A good way to attain better flow is to start with VSM. The value stream map will give employees an overall view of all warehouse activities, which allows them to suggest improvements in other areas, as well as their own. It is a good way for everyone to understand and agree on how the facility works and to come up with ideas for improvement. Display the current and future state maps in the warehouse so that employees are able to see previous improvements and take part in the ongoing effort.

To assess the operation using a value stream map, you need to involve the operators and supervisors, identify Lean improvements and *kaizens*, question every activity, treat the warehouse like a large staging area, develop justification as you go along, implement the Lean improvements using the VSM plan, and then start the cycle again.

5S or workplace organization ("a place for everything and every-thing in its place"), while seemingly simple, is a good place to start, and it is sometimes surprising how much time people waste searching for things.

Lean Tools in the Warehouse

In order to get any of this done, a team approach is necessary to identify the wastes in areas such as errors (receiving, put away, picking, loading, etc.), inventory inaccuracy, damage, safety, and lost time. Lean tools, such as problem solving (root cause, fishbone, etc.) and error proofing with standardized work (e.g., visual instructions on how to use a strapping machine or how to load/unload a truck) can also be helpful in the warehouse environment.

Pull systems using *kanbans* are a "natural" in a warehouse for everything from packing materials to forms, as well as product assembly and kitting.

If you do any value-added activities such as kitting or assembly, as many 3PLs do, then work cells might be appropriate to minimize labor and maximize the use of equipment and space. Line balancing is a tool that can be used in this type of situation not only in staffing, but also to ensure proper flow in the work cell.

There's even a place in the warehouse for total productive maintenance (TPM or equipment-related waste) as there's plenty of equipment (some automated) that might not be running as efficiently as possible (e.g., carousels, forklifts, hand trucks, strappers, etc.).

Lean Warehouse Examples

Menlo Logistics, a major 3PL provider, has not only implemented Lean at many of its facilities, but also uses it as a competitive weapon as can be seen at their Web site under "Lean Logistics" (www.con-way.com) where they point out the following areas where they look for waste:

▲ Mapping material flows— in studying material flows from raw material vendors to customer finished goods, we challenge each point at which material flow is stopped. Methods to speed up material flows include shipping ocean containers directly from Asia to inland regional warehouses, instead of transferring cargo to trucks at the ports of entry, and bypassing warehouses with large orders that travel directly from factory to customer.

▲ Keeping drivers and tractors moving—the interface between warehousing and transportation can often result in waste. We attack this problem by working with our carriers to drop trailers, freeing the driver and tractor to pick up a loaded one and keep moving. Through careful dock scheduling and synchronization of warehouse workflows, we can live-load shipments quickly, minimizing driver dwell time.

▲ Using milk runs—milk runs reduce transportation costs and build more consistency into an inbound supply network.

▲ Electronic data interchange (EDI)—Menlo Worldwide Logistics makes extensive use of EDI and RosettaNet to pass data among our supply chain partners. Communicating data electronically eliminates errors caused by manual data entry.

▲ Warehouse efficiency—Menlo Worldwide Logistics' team of industrial engineers designs warehouse layouts that streamline inbound and outbound flows, maximize labor efficiency, and deliver high space utilization. We employ techniques like dynamic slotting, cluster picking, task management and system-directed putaway to optimize labor and space efficiencies.

▲ Optimize transportation routes—Menlo Worldwide Logistics' LMS application optimizes each load to meet delivery dates with low-cost mode and carrier selection.

▲ Packaging optimization—We work with customers to explore the use of returnable containers for repetitive shipments to factories. For finished goods, we can study packaging sizes to uncover ways to increase pallet and trailer utilization. Small changes in carton sizes can facilitate better storage utilization and lower transportation costs. [www.con-way.com, 2011] Furthermore, Menlo emphasizes the use of mistake-proofing tools such as:

  ▼ Making it easy to do things right and making it hard to do things wrong

▼ Easy-to-read visual controls
▼ Radio-frequency devices coupled with bar-code technology
▼ System-directed cycle counting at our warehouses
▼ Utilization of Six Sigma and SPC
▼ ISO processes
▼ Electronic data interchange (EDI)
▼ Standardized processes
▼ Implement repeatable, standardized processes
▼ Establish one best way to perform each task
▼ Visual documentation of processes
▼ Correct any activity that causes rework, unnecessary adjustments or returns
▼ Organized workplace (5S) [www.con-way.com, 2011]

All of this results in benefits to their customers such as better service, lower costs, higher availability, higher customer satisfaction, and more reliable deliveries.

This isn't just "talk" either, as was pointed out by Gary Forger in *Menlo Gets Lean*. It describes how Menlo Logistics operates a 250,000 facility in Michigan that had recently shipped 8,000 orders in a 2 week period with no errors and according to a recent audit has inventory accuracy of 99.99 percent.

This site was the "pilot" site for their Lean program (along with a dozen other Lean warehouses at the time) with a goal to reduce the cycle time and increase productivity of various resources by eliminating waste. Menlo focuses their metrics on service, quality, delivery, cost, and employee morale. Warehouse operators work in 20-minute segments or small "batches" similar to what was mentioned earlier in this chapter. That maximizes flexibility and allows labor to help minimize response time to orders.

Items are slotted according to size and velocity, and workers are assigned certain aisles to keep neat and organized (and must sign off on a checklist).

Team leaders "own" their processes, supervisors and managers remove "road blocks," and bonuses of hourly team members are tied to metrics and improved processes (a real key to success, I believe). Besides weekly departmental meetings to discuss performance and improvement, every month a *kaizen* event is held in which as many as six workers concentrate on improving an operation. [Forger, 2005]

Peter Bradley in "The Skinny on Lean" stated that "Menlo Worldwide reports that warehouse productivity improved 32 percent between January and November last year, measured by gains in lines per hour. Defects, measured as the error rate, dropped by a whopping 44 percent. The on-time percentage for shipments was north of 99 percent in every one of those months, hitting 100 percent in eight of 11 months. And those involved think they can do more." [Bradley, 2006]

Another major 3PL player, Ryder Logistics, highlights their "Five LEAN Guiding Principles" on their Web site (www.ryder.com), which "provide the foundation for operation excellence, continuous improvement and supply chain efficiency." The guiding principles are: people involvement, built-in quality, standardization, short lead time, and continuous improvement.

Ryder also mentions using a variety of Lean tools in their business, such as workplace organization, visual management, work cells, standardized work and even a Lean Academy. They not only include Lean applications in the warehouse operations, but start with determining the optimal distribution network design, which can significantly reduce waste in the overall supply chain network. They take this very seriously.

A Ryder case at the site describing how they took over Whirlpool's service operation shows how continuous improvement activities reduced costs, improved shipment accuracy and order cycle time, and boosted overall efficiency. The first thing they determined was that it was more efficient to consolidate Whirlpool's various service facilities into one location. After that, they implemented a variety of continuous improvement efforts, including the creation of an inventory profile that identified the best storage location for each part to improve efficiency of order picking, improved workflow processes leading to more efficient use of labor, and collaboratively enhanced the existing WMS system that enabled them to streamline the operation further.

Ryder then keeps track of five key performance indicators (KPIs) on a monthly basis (the first four of which are also measures of waste): shipment accuracy, inventory accuracy, order fill rates, order cycle times, and budget performance. [www.ryder.com, 2011]

The point of these examples is to show that not only is the warehouse an ideal place to start a Lean supply chain and logistics journey, but that it can give you real results and a competitive advantage in the marketplace.

CHAPTER 9

Lean Global Supply Chain and Logistics: The Long and Winding Road

The trend in global outsourcing, which started in the 1980s and continues today, has made our supply chains much more complex and challenging. There are, of course, other reasons to globalize, such as private labeling strategies or to take advantage of the rise of business and consumer markets in China, India, and other markets around the world. All of this adds time and distance to your supply chain, which can result in higher cost, complexity, and risk.

The risks can be economic (the "great recession"), environmental (oil prices, earthquakes, and tsunamis), or political (war, revolts, and trade barriers) and, as a result, you can understand that there might also be opportunities for a leaner and more agile supply chain and logistics process.

The Logistics of a Global Supply Chain

In fact, an IBM Global Services white paper, "Five Reasons Why Global Logistics Is Moving from the Basement to the Boardroom," found that as global sourcing has become more prevalent, logistics planning, specifically on a global scale, has risen in importance. As a result, executives look to international logistics to gain a strategic competitive advantage and also to obtain value—both areas where Lean can help. The white paper points out that executives have noticed and now focus on inventory, working capital, service, and Lean Manufacturing issues. The "five reasons" given for global logistics increasing importance are:

1. **Managing rising logistics costs**—at 10 percent of total costs now, and as they increase, they diminish some of the advantages of global sourcing.

2. **Lean Manufacturing**—logistics must support the goals of a Lean Manufacturing program. While inventories have decreased as a result of Lean, the longer lead times and transit times of global sourcing put added pressure on the logistics function to also be agile and Lean.

3. **Operating as an on-demand business**—as more companies move to a "mass customization" requiring a very flexible, collaborative type of approach, the logistics operations must follow suit.

4. **Cross-functional sourcing decisions**—the increase in global sourcing and outsourcing requires detailed logistics information. Changing a vendor or manufacturing location has potential huge impacts on logistics cost and performance. The same can be said about tapping into new markets.

5. **Supply chain management**—detailed, timely, and accurate information is needed to make the right transactional, planning, and scheduling decisions for all areas of the business. This information includes access to optimized logistics costs, predictable delivery times, visibility, carrier performance, and automated data and analytics. Many times, companies ship via air freight thinking they will get better transit times for premium shipping fees. However, delays in customs or other areas can reduce or eliminate the benefits of premium air freight services. You need to look at the entire journey as one "value stream" to help make the best decisions for your company. [Taylor, 2006]

All of these areas can be better managed and improved using Lean concepts.

Value Stream Mapping to Identify Waste

As we know, the supply chain is a global network that is used to deliver products and services from raw materials to end customers through the flow of information and physical distribution, and as a result, there are plenty of opportunities to identify wastes. The most efficient way to identify these wastes is by VSM. But where are the best places to look for waste in global supply chain and logistics process?

Areas of Potential Waste in the Global Supply Chain and Logistics Network

Transportation and inventory play vital roles in domestic and international supply chain management. Transportation cost can be drastically reduced if the frequencies of transportation are reduced or minimized. In most systems, transportation costs are volume dependent and by utilizing efficiently the trucks' capacities, transportation costs are minimized. Economies of scale will result from system-wide safety stock being reduced whenever decentralized inventories are centralized into fewer locations. As previously pointed out, inventory levels are often the cause of increased costs and cover waste in companies.

When dealing with global logistics, an extraordinary number of parties are involved in the process. They can include customs, freight forwarders, banks, ports, transportation (rail, truck, and ocean), etc.

As there can be as many as 15 parties or more involved, the collection of data for a value stream map can be quite a challenge. Tom Craig, president of LTD Management (www.ltdmgmt.com), points out in his article "International Lean Logistics—Beyond the Four Walls" that, "a supplier in Shanghai whose key component comes from Thailand must participate actively in the mapping since all this is part of the process. This is not an option. or look at a customs broker who does not directly touch the product or the shipping container. He acts with the information and documentation to facilitate the movement of the product. But the linkage among the importer, customs broker, ocean carrier/air forwarder and delivering rail or trucker can create waste, by adding times and by stopping product flow."

Mr. Craig demonstrates in his "current state" value stream map of a typical international logistics move (Fig. 9.1), that the entire cycle can take up to 126 days when including the demand planning, supplier, and logistics performance functions and that, in many cases, companies tend

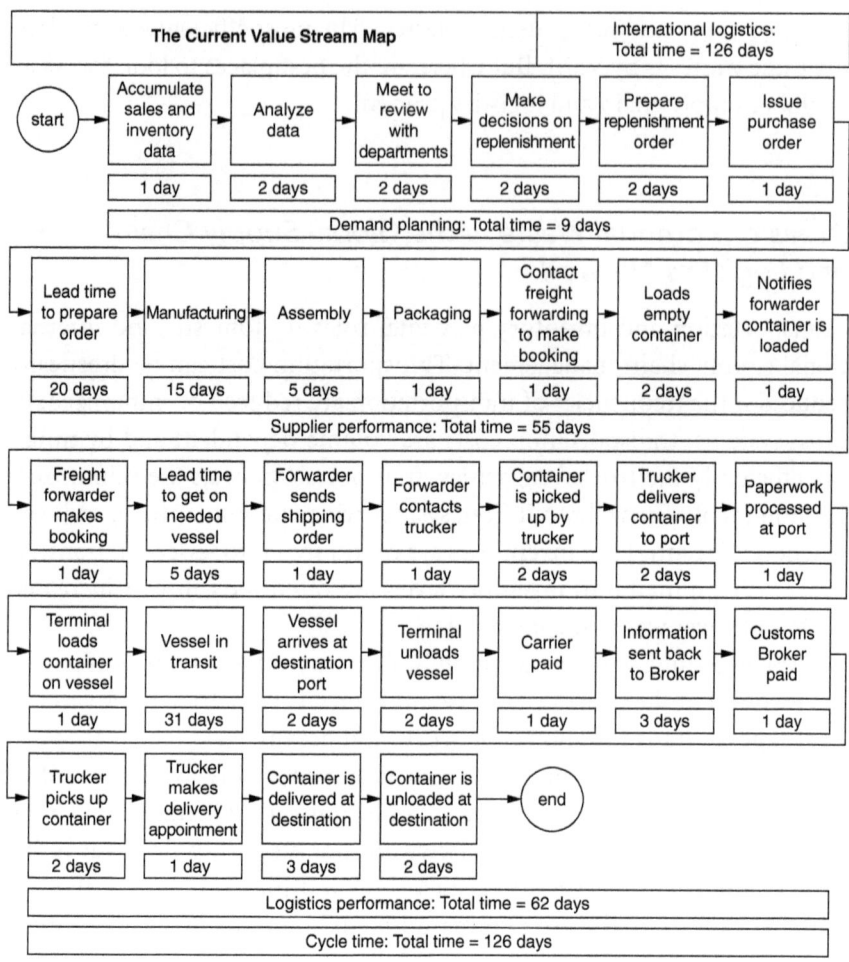

Figure 9.1   International logistics current state value stream map.

Source: "International Lean Logistics—Beyond the Four Walls" by Tom Craig, LTD Management, 2011.

to push most of the waste onto third parties. It takes real collaboration to reduce this waste.

A thorough analysis points out areas of potential waste in all areas of the current process such as the fact that:

…more than 25% of purchase orders are not shipped as planned or are not delivered as planned. This significant statistic presents a real opportunity to reduce waste. Supplier performance and supplier lead times are important areas for potential waste reduction and process improvement.

Also, the distribution network may be outdated. It may have been built years before with different store or customer configurations, different products, and other topics. It may have been built when the focus was on storing inventory in warehouses, unlike now when inventory velocity is emphasized. Touching the product to store it often adds only time—a waste result, not value (see map at bottom of facing page).

Bypassing warehouses with cross dock or other transfer facilities at ports can remove time and inventory. Supply chain execution technology can give visibility from the purchase order through to delivery order. It can provide the way to allocate product in transit. Making this part of the new process reduces two key wastes—time and inventory.

Global supply chain management has significant "built-in" time because of the distance involved. This runs counter to domestic supply chains. The extended time can, in turn, create uncertainty and the need for many companies to build and carry additional inventories. Yet time and inventory are two areas of waste for lean to improve. So, Lean international logistics faces an additional challenge because of its inherent scope and the impact throughout the supply chain, especially within the company.

As a result of this type of analysis and collaboration, the global supply chain can be looked at as one process, not many different ones (i.e., purchasing, transportation, customs broker, freight forwarders, etc.) to ensure proper flow and minimal waste. In the previous example, the total cycle time can be reduced to as little as 97 days through applying Lean concepts (see Fig. 9.2). [Craig, 2011]

The use of simulation tools in conjunction with VSM can also help identify gaps between expectations and potential outcomes before implementation.

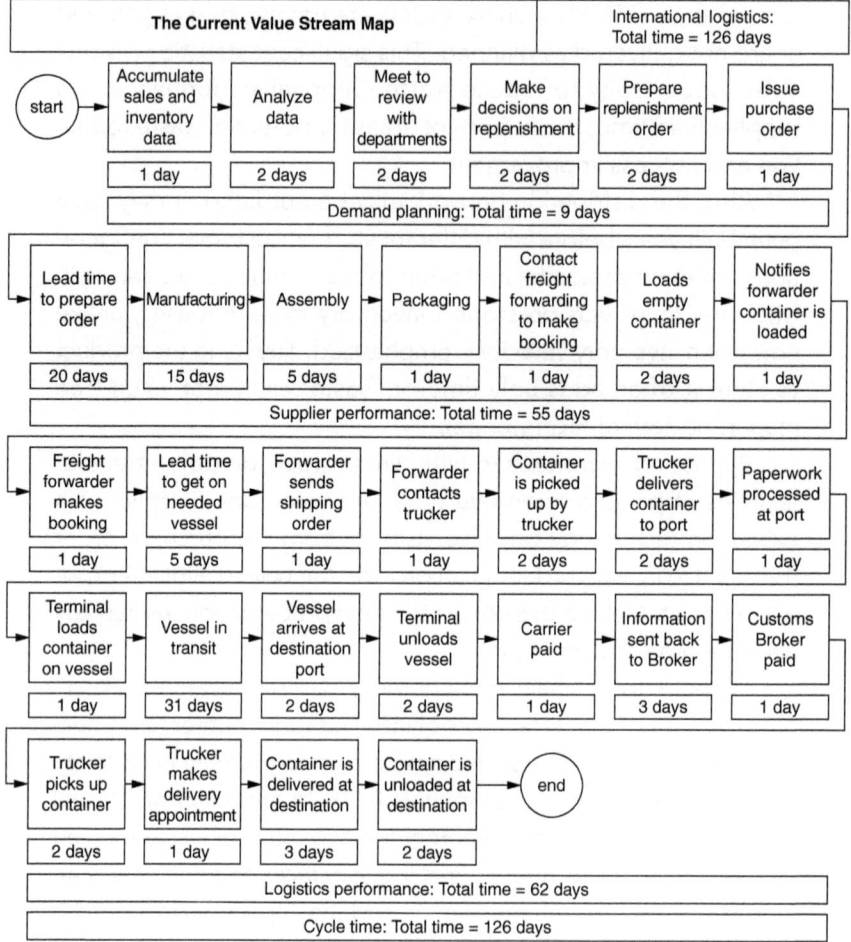

Source: "International Lean Logistics—Beyond the Four Walls" by Tom Craig, LTD Management, 2011.

Figure 9.2 International logistics future state value stream map.

Areas to Reduce Waste

A 2006 survey from McKinsey and the U.S. Chamber of Commerce of Western companies importing products from China entitled "The Challenges in Chinese Procurement" found that the most of respondents felt that they were behind their competitors in such outsourcing areas as quality, sourcing cycle time, and other key supply chain metrics. [Hexter and Narayanan, 2006]

So that leads to the question of why this occurs. According to a 2006 *Supply Chain Digest* white paper entitled "The 10 Keys to Global Logistics Excellence," there is a tendency to overestimate the savings in the first place or perhaps the savings were lost because of poor execution. [www.scdigest.com, 2006]

The cost could be miscalculated because of a number of things, including higher-than-anticipated transportation costs or more buffer inventory required because of longer replenishment times.

Poor execution can easily occur in the world of global logistics. The expertise required to handle the various languages, currencies, duties, tariffs, etc. can be overwhelming for even someone with a lot of experience in this area. Lack of appropriate technology can also be a detriment in this area.

Keys to Global Logistics Excellence

Most if not all of the 10 keys to global logistics excellence mentioned in the 2006 *Supply Chain Digest* white paper pertain to Lean and should be considered when looking for waste in this area. They are:

1. **Total delivered cost management**—There are a plethora of logistics costs when sourcing and shipping globally. They include domestic and international transportation and distribution costs by product and route. You also need to include duties, tariffs, and other customs costs. Having automated systems to track and measure performance this information helps, of course.
2. **Global logistics process automation**—The more automation the better, especially in terms of visibility. Also, a lack of automation results in too much time being spent on manual activities instead of planning. Global logistics execution is complex with as many as 25+ handoffs in one shipment with multiple parties. Many ERP and logistics systems do not support global logistics as well as is required by the users.
3. **End-to-end visibility**—Global visibility and event notification are critical to minimizing waste in your supply chain. This is not only useful for tracking but also spotting and reacting efficiently to exceptions. Timeliness and accuracy are extremely important when managing your global supply chain. Examples include EDI with ocean carriers, web portals to 3PL providers, and custom brokers. This allows companies to make the most of their internal labor and other

global overhead resources while being in control of their outsourced global supply chain.

4. **Supplier portals and *advance ship notice* (ASN) capabilities** —Even though you may have adequate transportation visibility, you may still be lacking the status at the foreign factory or the details of what is on each container. That is where ASNs and supplier portals help you to get early notification for this type of information, thus giving the incoming distribution center advance notice and helping to plan inventory better (and thus reduce waste). Many companies still receive this information via fax, and in many cases they cannot be 100 percent sure of accuracy until opening the container.

5. **Total product identification and regulatory compliance**—Security, import/export restrictions, customs, and other safety and regulatory requirements hinder the continuous flow of the global supply chain both slowing and adding cost to it. Technology can assist in this area, such as the use of RFID systems and software that works with denied screening and other regulatory requirements.

6. **Dynamic routing**—Even on repeated routes, for both cost and agility, shippers now look for different combinations of carriers, routes, and freight forwarders similar to how it has been done domestically for years. They not only get lower rates, but can have a more agile supply chain with more dynamic alternatives (such as direct ship, drop ship, cross dock, etc.). This can also lessen risk, such as the recent earthquake and tsunami in Japan in which it would have been ideal to have a backup supplier of semiconductors with all the logistics costs and processes known in advance.

7. **Variability management**—As we know, variability is a contributor to waste and has a huge impact on both inventory and customer service, and in the area of international delivery times, it is even more prevalent. By using performance measurements and tools, such as root cause analysis, supply chain managers can attempt to minimize the variability for better, more predictable global supply chain and logistics performance.

8. **Integrated international and domestic workflow**—Until recently, there were not transportation management systems (TMS) that integrated domestic and international logistics, almost forcing companies to look at them as two separate shipments. There is still

a lot of room for improvement on this front, that will eventually allow for more centralized international logistics similar to what has existed on the domestic side for years.

9. **Integrated planning and execution flow**—The idea here is to attempt to have technology on the global logistics side that follows the planning and executive process (i.e., automated and integrated). Ideally, this should be real-time or close to it, and as was pointed out earlier, much of the international freight movement data is still collected manually. This is truly the ultimate way to make sure your global supply chain and logistics function is Lean.

10. **Financial supply chain management**—It is especially critical on the global logistics side that financial processes for things like letters of credit, financial settlement, and other import/export documentation do not interfere with the flow of goods. So it is important to have both internal personnel skilled at this, as well as partners and software vendors, to make sure that international transportation planning and execution are integrated and as efficient as possible. [www.scdigest.com, 2006]

Addressing Wastes in the Global Supply Chain

Tom Craig, in his "International Logistics—Beyond the Four Walls" article points out that there are many other areas to look for waste that are specific to global supply chain and logistics and include:

▲ Use technology to manage supplier performance and to integrate the movement of information among and between all parties.

▲ Design a process that is lean and includes all parties and that differentiates among different commodities and products and among different customers.

▲ Collaborate with key suppliers and logistics providers.

▲ Link demand and demand planning with replenishment and buying.

▲ Reduce the number of suppliers and logistics service providers to streamline the supply chain, without sacrificing results.

▲ Focus on supplier performance; control the supply chain at the international source. The offshore supply chain begins with the purchase order; transportation is a derivative of the purchase order and of supplier performance.

▲ Understand transport differences and options such as ocean carriers offering different transit times, different sailing schedules, different destination ports and different canals to the East Coast (Panama Canal versus Suez Canal).

▲ Align your financial supply chain with your trade supply chain. These two chains involve different sets of players with differing objectives and practices.

▲ Use a 4PL or 3PL to manage your offshore supply chain. Work with a supply chain service provider that understands the total supply chain complexity and operation. His interest should be your supply chain, including your suppliers and purchase orders, not just your freight. The firm should use process, technology and people to do this. The people should be located in the same country and locality as your suppliers. [Craig, 2011]

According to international supply chain consultant David Sardar of Cross Country Consulting (www.crosscountryconsulting.biz), areas to look for waste in international supply chain and logistics include "order processing, processing [of] all export documentation, clearing customs, and transportation. Partnering with a broker that knows the industry and your business is critical. Very few, if any, brokers are good for all locations. Local brokers have contacts in certain ports and this helps with their knowledge of what causes delays in certain ports."

Furthermore, Mr. Sardar has found that "minimizing time and costs from final inspection to the point of sale or use is the goal. To do this it would start from how quickly suppliers process orders, pick, pack, ship, clear customs and get it to the destination. Customs and getting it on the best ship or plane and to the best destination port or airport is the key. In making our supply chain leaner we were able to eliminate a few non-value added steps and this reduced our total cycle time as well as reduced costs." [Sardar, 2011]

Global supply chain and logistics is a complex process. As described in this chapter, there are plenty of opportunities to look for activities that are non-value-added from the customers' viewpoint using tools such as VSM and using a continuous improvement process to eliminate or at least minimize them.

CHAPTER 10

Keys to Success: The Patient Gardener

Over the years, many fads and buzzwords have come and gone. In many cases, it is because they were not really great ideas, but in other cases, it was because there was not the proper environment necessary for them to take root and grow.

If you look at how companies are run overseas, you will realize that they tend to take a longer-term view of the world than Americans. They also focus more on training, as was pointed out earlier the book (U.S. firms average 7 hours of training per year, while other advanced countries average hundreds of hours per year).

The Lean transformation process is truly a journey that can take a long time and requires patience and support. Companies that expect huge short-term results are usually disappointed and tend to give up, while those that stick with it can attain some significant, long-term results, including lower costs, shorter cycle times, inventory reductions, increased capacity, reduced defects and rework, and greater employee morale.

Key Success Factors

There are a number of *key success factors* (KSF) critical for a successful Lean journey.

Lean Training

First of all, it is important to train the entire organization to ensure that everyone understands the Lean philosophy, as well as has an understanding of all of the concepts and tools. It can be useful to have an outside trainer and facilitator to come in to lead *kaizen* events, especially in the early stages. As they used to say about consultants, "They borrow your watch to tell you what time it is." Recently, I delivered some 5S training and *kaizen* events at the truck maintenance facility at a 1.5-million square-foot distribution center for a major toy retailer. Due to rapid growth in the fleet and limited space, the storage area of the facility had become quite a mess. While they were aware of this, it took someone from the outside to help them to look at things differently and understand the benefits of workplace organization. The end result was more storage space, which was desperately needed (and it was much more organized so it was easier to find spare parts and supplies).

Management Support

It sounds simple, but no type of change can be successful without top management actively driving and supporting the change with strong leadership. Often, when companies do employee training using grants, management seems to look at it as "free" training, without considering the potential long-term commitment required to get the potential benefits of the training. That type of thinking is rarely successful for anything other than short-term results. Management has to be in it for the long haul, otherwise it is doomed to fail. In fact, it is a good idea for your overall company business strategy to include Lean.

There has to be a commitment from *everyone* in the organization to make it work, not just management. Lean and continuous improvement in general should become part of everyone's job, from the job description to the review and reward system. A small amount of time should be set aside during every meeting to discuss Lean projects and progress. Everyone should be encouraged to question the status quo.

Lean Structure

It is always a good idea to designate a *Lean champion* to spearhead continuous improvement activities. In many cases, those who really believe in the benefits of Lean hire full-time Lean champions. When doing so, you should try to find a good, experienced change agent as the champion. The more structure you have to accomplish your Lean goals, the better chance for success. Some companies also assign *Lean coordinators* and even *subject matter experts* (SMEs). The coordinators are responsible for various *kaizen* events and the SMEs may be trained to train and manage for specific types of events (e.g., 5S, VSM, etc.).

It is usually the Lean champion's responsibility to create a *kaizen* agenda and communicate it to operators and others in the organization, who are members of the empowered teams.

To get started, it is usually best to pick an area that is visible, yet manageable in size and one that can be dramatically improved, such as a small supply storage or work area with a lot of traffic. This way, the organization can see that you are both serious about Lean and can see how it can make a difference. As mentioned before, after general Lean overview training to everyone, some companies go right to VSM, and others start with 5S events. Which way a particular company goes depends on its size and the industry it is in, among other things. In any case, it is important to start mapping value streams to identify opportunities and begin as soon as possible with an important and visible activity. While doing this mapping, it is always helpful to benchmark or compare data within your industry and outside your industry to help prioritize improvements.

It is important to become as lean as possible within your company's "four walls," then expand to customers and suppliers. In many cases, if you do not eventually partner with suppliers and major customers on the Lean journey, the full benefit of Lean is never reached and in some cases, you are just passing your inefficiencies on to suppliers, making them less efficient. One Lean client we had was partially funded by a major (large) customer who had been through a Lean journey themselves. They had reached the point where they wanted their major suppliers to become more efficient as well. There were however "strings attached"—the supplier was responsible for improving specific metrics such as on-time delivery, lead times, etc. That way it was a win-win situation for both the customer and supplier.

Teamwork and Lean

Teamwork is fundamental to compete in today's global environment in general and is fundamental to Lean in particular. Teams help support a process that defines and solves problems using Lean tools.

These days, most employees have been on some kind of team, whether it is a management team, quality circle, new product team, etc. It is pretty much the "way of the (business) world" these days. In fact, it seems that teams are more the norm than the exception in today's environment.

Making Teamwork Happen

To make teamwork happen, executives must communicate the clear expectation that teamwork and collaboration are expected from the group. Management must talk about and identify the value of a teamwork culture and reward and recognize it. Over time, stories and folklore develop that people discuss within the company that helps to emphasize teamwork.

Most successful teams have a large degree of employee empowerment or the authority to make decisions on their own. In many cases, teams are self-directed, which gives the members this added sense of empowerment. In order to ensure success, management must provide adequate support, training, and establish clear objectives and goals. There is always the "what's in it for me" question, so the idea of financial and non-financial rewards are important. From a simple "pat on the back" or treating the team to lunch to actual financial rewards can be useful and meaningful.

In general, it can sometimes be difficult for managers to give up control to the team, but as long as they make sure that there are adequate controls in place, it can be successful.

When teams are successful, organizations and employees can benefit from an improved quality of work life, increased motivation, improved satisfaction, better quality and productivity, and lower turnover, among other things.

However, on the other hand, there are higher costs involved as a result of greater wages, training, and capital costs needed to be successful.

Sales and Operations Planning (S&OP)

We will delve more into teams and *kaizen* events in the next chapter when we discuss VSM and how teams and *kaizen* events are necessary to get to the "future state." In the meantime, we will discuss the topic of *sales and operations planning* (S&OP), a function that is also team-based, but rarely discussed in the context of Lean Enterprise.

S&OP Defined

In its simplest terms, S&OP is a way for a business to ensure that supply can match demand, at least on the aggregate. As discussed previously, Lean teams plan and execute on a shop-floor level, but S&OP can be a great tool to make the connection between the *kaizen* event goals and objectives and corporate ones. As we know, inventory is one of the eight wastes and covers variability in a process. Through the use of S&OP, inventory can be directly controlled. In general, there are two general ways to reduce inventory: (1) more accurate forecasts and (2) shorter cycle times. The S&OP process attempts to improve and control both of these.

S&OP, also called aggregate planning, is a process in which executive-level management regularly meets and reviews projections for demand, supply, and the resulting financial impact. S&OP is a decision-making process that makes certain that tactical plans in every business area coincide with the company's business plan. The net result of the S&OP process is that a single operating plan is created that allocates company resources.

S&OP increases teamwork between departments and helps to align your operational plan with your strategic plan. It is a process in which various targets are set (e.g., forecast accuracy, inventory turns, etc.), and progress against the strategic and operational plans are reviewed in a series of meetings.

The objective of S&OP is to have consensus on a single operating plan that meets forecasted demand while minimizing cost over the planning period. It should allocate people, capacity, materials, and time at the least possible cost, while ensuring the highest customer service possible.

Supply and Demand Options

In order to meet predicted demand, there are capacity options available, which include the adjustment of production rates, labor, and inventory levels, as well as the use of overtime, part-time workers, and subcontracting.

Alternatively, there are also some demand options as well. These include influencing demand, backordering, counter seasonal product offerings, and service mixing.

These approaches can be implemented to varying degrees by using what are commonly known as chase, level, or mixed strategies.

A chase strategy allows production to meet demand for each period by adjusting production or labor rates. This works wells in the service industry, but in goods it can cause major headaches from constantly laying off and rehiring and training personnel (as well as the added costs as a result).

The level strategy basically uses inventory as a buffer to keep staffing at consistent levels. There are downsides to this such as carrying way too much inventory, and as inventory is made much earlier than needed, there may be too much of the wrong items and not enough of the right ones!

Many companies, use a combination of the various capacity options mentioned previously, which is called a mixed strategy.

The S&OP Process

The series of meetings prior to the final S&OP executive-level meeting are:

1. A demand planning cross-functional meeting, at which forecasts are reviewed with a team that includes (at the very least): operations, sales, marketing, and finance. Typically, forecasts have been generated statistically and aggregated in a format that everyone can understand and confirm (e.g., sales might want to see forecasts and history by customer in sales dollars).

2. A supply planning cross-functional meeting at which agreed-upon forecasts have been "netted" against current on-hand inventory levels while factoring in information such as safety stock targets, lot sizes, and capacity constraints to create production/purchasing plans. Again, this data will usually be reviewed in the "aggregate" by product family in units, for example.

3. A pre-S&OP meeting, at which data from the first demand and supply meetings are reviewed by department heads to ensure that consensus has been reached.

As the logistics manager at a major household products company, we used this type of process (even before it was called S&OP), to accomplish aggregate production planning for the various company-owned and -contracted manufacturing sites. We would always put pressure on the plants to cycle through our products more often, which was a challenge. In the past, they were used to scheduling production on a monthly basis with some priorities. We used the ABC method to push the plants to run the A items every two weeks (and eventually weekly), B items monthly (eventually biweekly), and so on. It took a lot of time and continued pressure to get the plant manager behind us as is typically the case, changeover costs were the issue, resulting in larger-than-needed batch sizes.

Simultaneously, we put in place a collaborative forecasting process using a statistical forecast as a "base" at the item level (actually at the SKU level—item at a distribution center), which was aggregated in various units of measure and levels to share with marketing, sales, and finance among others. Forecast accuracy targets and measurement (by ABC code) were crucial to this process's success.

We would have a monthly process where we would first come up with the most accurate collaborative forecast possible and then run our production plans to match supply with demand, which was then reviewed by a management team.

Over time, the company was able to significantly reduce inventory levels while improving customer service directly as a result of this process.

S&OP increases teamwork between cross-functional areas. It puts your operational plan in line with your business plan. S&OP is a formal process in which targets are set, and progress against the business plan is reviewed.

S&OP and Lean

From a Lean perspective, a robust S&OP process acts as both a planning *and* control method at an executive-management level as various metrics indicating the level of waste, such as forecast accuracy, inventory turns, and on-time and complete shipments, to name a few, are benchmarked externally to set objectives (as well as matching the company's strategic plan) and also measured to know when things are in or out of control.

According to an Aberdeen Group study in 2010, entitled "S&OP—Strategies for Managing Complexities with Global Supply Chains" by Nari Viswanthan, four key performance criteria distinguish best-in-class in terms of S&OP. They were:

1. Forecast accuracy
2. Perfect orders delivered complete and on time
3. Cash-to-cash cycle
4. Gross profit margin [Viswanthan, 2010]

All of these, especially the first three, are measurements of how Lean a company is in that the lower the score, the more "variability" your system has which leads to many of the wastes we have discussed.

Working Together

Dougherty and Gray in their 2006 book *Sales and Operations Planning—Best Practices*, point out that, in a study of 13 best practice companies, Lean and S&OP help each other. According to Malcolm Jaggard, director of supply chain management, AGFA US Healthcare, at one of their best practice clients, "continuous improvement is embedded in the S&OP process, and continuous improvement cannot be maximized without S&OP."[Dougherty and Gray, 2006]

Furthermore, the authors point out that:

If you create a manufacturing environment where material flows with minimum waste (Lean), but you can't predict capacity and material availability problems in enough time to avoid them (S&OP), you will inevitably revert to firefighting, finger-pointing and poor results. Similarly, if you do an excellent job of future planning but have poor flows, you can almost count on higher inventory levels, longer lead times, and lower profitability... Traditionally lean manufacturing has been stronger on workplace management; S&OP on decision-making for the future. The tools and methods of lean manufacturing have tended to look most closely at the plant, and its immediate customers and suppliers, mostly over a short horizon. This leads to improvements like: "shorter, quicker, fewer, lower cost, more flexible, and better aligned." S&OP provides distance vision—providing the ability to predict capacity and material availability problems before they become crises, to identify market issues while they are still opportunities, and to prioritize improvements in a way that will create the most favorable results. [Dougherty and Gray, 2006]

Tomorrow's supply chain will be driven, among other things, by effective S&OP, which allows for effective supply chain planning, balances new and current products and services, employs timely and effective replenishment, enables timely success/failure measurement, and, with the help of technology, can generate data/analysis/correction.

When S&OP is implemented in a company that is focused on a team-based continuous improvement process, from top to bottom, success can be ensured.

CHAPTER 11

Getting Started: Lean Forward

So far, we have discussed many of the concepts and possible applications in supply chain and logistics management. This leads to the question "where do we start the Lean journey?"

Lean Opportunity Assessment

In many cases, the best place to start your journey is to perform a *Lean opportunity assessment* (LOA) to help identify the potential for improvement in your organization (a useful template, Appendix B, accompanies this book). In this type of assessment, you should identify and analyze various aspects of your company from a Lean perspective and rate them against best practices.

In our template (and there are many variations on this type of assessment), we look at various key performance areas and rate them anywhere from "traditional supply chain and logistics" (i.e., "beginner") up to "world-class supply chain and logistics management."

The areas recommended for evaluation are:

- Internal communication
- Visual systems and workplace organization
- Operator flexibility
- Continuous improvement
- Mistake proofing
- Quick changeover
- Quality
- Supply chain
- Balanced production
- Total productive maintenance

▲ Pull systems
▲ Standard work
▲ Engineering
▲ Performance measurement
▲ Customer communication

Users then rates their performance in each category on a scale from 1 to 5 and compare to Lean best practices to see where general opportunities for improvement exist.

Value Stream Mapping

While an LOA is a good first step, in order to actually identify, plan for, and make the improvements, you need a road map. The perfect road map for Lean is what is called a value stream map (VSM).

A *value stream* is the set of all actions (both value-added and non-value-added) needed to deliver a specific product or service from raw material through to the customer.

A VSM is typically the initial step your company should take in creating an overall Lean initiative plan. By developing a visual map of the value stream, it allows everyone to fully understand and agree on how value is produced and where waste occurs. It should be noted that in many cases, after company-wide general Lean introduction training, many companies start instead (or at the same time), with a 5S program, which as mentioned before is a good foundation concept and useful in transforming the company's culture from the bottom up.

Historically, businesses used *flowcharts,* which show the movement of materials, *time function maps,* which show flows and time frames, and *process charts,* which use symbols to show important activities. While these were adequate, they were typically done by individuals (often consultants) who interviewed various parties in a process in order to document it.

In many cases, this type of process mapping is used to enable *business process reengineering* (BPR), which is a total rethinking and redesign of a process. It was very much in vogue in the 1980s through 1990s, when companies were dramatically consolidating and outsourcing, resulting in large numbers of plant closings and layoffs in manufacturing.

This is in contrast to the concept of incremental change or continuous improvement, as it does not give any indication as to whether individual activities are value-added or non-value-added, especially from the viewpoint of the customer. That is one of the many improvements when using VSM, along with the fact that some data is also gathered to help to quantify the waste in an *entire* value stream from customer to supplier.

Value Stream Mapping Defined

VSM is a team-based approach to mapping a value stream or process from beginning to end. It visually (and numerically) breaks the process down into value-added and non-value-added steps from the viewpoint of the customer. Typically, you should start at the customer end and work your way back to the supply end. It is best to have a diverse team of employees (no more than 8 to 10) who are usually supervisory and management level (although front-line employees will be involved in the mapping process as well). It should take no longer than 2 days; 1 day for training and the current state map and the second day (usually a week or so later) to create the future state map. The entire process, including implementation, should be completed in less than 6 months at most (if not sooner).

The team first maps the current state, while at the same time gathering ideas and input for an improved, future state. The maps are drawn using standard symbols (easily found on the Internet) showing the flow of materials (shown at the bottom of the map) and information (at the top of the map). It should typically take no more than 1 day to map a processes current state, and this is really a "10,000-foot" level view so it should not be too detailed (but information should be validated).

The future state map, using data contained in the first, is the same value stream with any waste, defects, and failures eliminated.

It is usually best to put some time in between doing the current and future state maps to give time for the ideas to germinate.

Before starting on the current state map, it is important to determine:

1. The value stream you are mapping
2. The product families that will be included
3. The takt time for the selected product family
4. The value stream manager
5. Your goals and objectives

Value Stream Mapping Benefits

There are many benefits to VSM, especially when compared to the other methods described in the previous section.

A VSM shows the various connections between activities, information, and material flow that can impact the lead time of your value stream, allowing you to separate value-added activities from non-value-added activities and then measure their lead time.

It can also help employees understand your company's entire value stream rather than just a single part of it, as people tend to get tunnel vision or a "silo mentality," which is counterproductive when trying to implement Lean. VSM also provides a way for employees to easily identify and eliminate areas of waste.

At the same time, VSM can improve the decision-making process of employees by helping team members understand your company's current practices and future plans.

Other benefits include:

▲ Establishes priorities for improvement efforts
▲ Focuses on no-cost or expenseable improvements
▲ Provides a common language to talk about the processes
▲ Is based on objective information
▲ Forms the basis of an implementation plan

Value Stream and Product Family

Usually, the value stream selected to map for the first time, is a process within the four walls of your company. However, in the case of supply chain and logistics management, this may not always be the case as the supply chain connects us with our suppliers and customers. We may be mapping a process in our warehouse or distribution center, or a process such as purchasing and/or transportation, which is directly related with our suppliers.

In any case, it is important to first narrow the scope of our VSM to a specific product family. A product family consists of products or services that have mostly the same steps. That does not mean that all the products or services have to have all of the same steps (just mostly the same steps!).

Takt Time

As discussed earlier, takt time, which is the demand rate for a product or service family, shows the pace for each activity in the value stream and is useful in identifying bottlenecks, which constrain or limit capacity. In a VSM, the takt time is usually calculated in minutes or seconds (i.e., the need to ship so many units every so many minutes or seconds).

Value Stream Manager

Every team needs a quarterback and each value stream needs a value stream manager. For product ownership beyond functions, you should assign responsibility for the future state mapping and implementing lean value streams to line managers with the capability to make changes happen across functional and departmental boundaries. Value stream managers should make their progress reports to the top manager on the site.

Goals and Objectives

It is a good idea to create a "steering committee" to manage and support the Lean journey. This is especially true in the case of VSM, as it sets the direction of the company and drives individual *kaizen* events.

It is also critical that the goals for the VSM and subsequent *kaizen* events tie to the company's strategic and operational goals. This usually is not too hard to do as long as the company has established these goals. Typically, they can include metrics covering quality, manufacturing, supply chain and logistics, and customer service to name a few. Specifically, they may measure things like inventory turns, on time and accurate shipments, cases/hour handled, etc.

In a more general way, when setting goals and objectives, the team should be asking questions like:

▲ What does the customer want in terms of the cost, service, and quality of our products/services, and which objectives and goals established by our company meet these market needs?
▲ What processes greatly impact the performance of these products and services?
▲ Who needs to support this effort?

Steps to Creating a Current State Value Stream Map

While there are standard "icons" that are used in VSM (easily found on the Internet), there really is no "standard" step-by-step methodology to the actual mapping process. In general, you should:

1. Have individual team members do a "quick and dirty" map of the major processes in a value stream by themselves and then as a group, compare maps, and reach a consensus. This should typically just include the individual processes (from customer to supplier) to make sure that none were missed. It is always useful to draw the major activities on a whiteboard once they have been agreed to at this point.

2. Assemble data collection forms (to be discussed shortly), pencils, "yellow stickies," and a stopwatch for collecting data.

3. Select a product family (or service) to map. Conduct a quick tour of the value stream to view the end-to-end material and information flows, making sure you have identified all the component flows (it is usually best to start in shipping and work your way back). Always let operators know in advance about the tour so that they don't work any faster or differently than normal. During this time, it is important to observe and ask questions and to start thinking about the "future state." These ideas should be captured on the "yellow stickies."

4. Identify a representative customer(s) of the product and gather data about typical order quantities, delivery frequencies, and number of product variations (will be used to develop takt time).

5. Assign team members individual value stream activities for which to collect data using the data collection form.

6. Begin mapping the detailed value stream, (using Lean icons), starting with customer requirements and going through the major production activities. The result is current state value stream map (information flow on top, material-flow in the middle, and a lead-time chart at the bottom of the VSM, showing value-added and non-value-added production lead times).

7. Review the map with all the employees who work in the value stream you have mapped to ensure accuracy (see Fig. 11.1 for current state VSM example).

8. Review the map with upper management before proceeding with the future state mapping process.

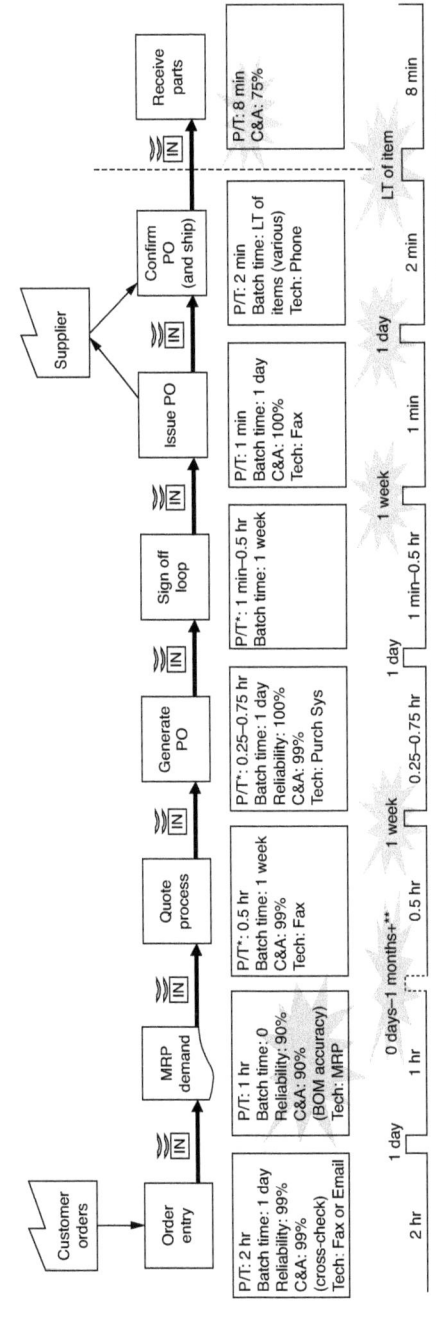

Figure 11.1 Current state value stream map example.

* depends on $ value

** parts, electrical components, etc. MRP req not being consistently acted upon.
new product requirement told by Eng not to buy, but there are some legitimate ones as well
(maybe be due to quantity low and told to wait by O&R. wrong lead time which is too long.

Data Collection

When doing VSM in supply chain and logistics management, in some cases, it may be more like a "traditional" VSM in manufacturing, where you look at the inventory that is built up between processes as non-value-added and convert the quantities to days or even minutes using the takt time. However, in both the office and warehouse, it is typically more important to look at inventory "dwell" time between processes as a better indication of waste.

In a warehouse or distribution center (DC) it is important to see how long inventory sits between receiving and put away, picking and staging, staging and shipping, etc. It is always a good thing to also measure the amount of inventory between steps, but in many cases, the DC may have no control over the amount of inventory (a VSM and *kaizen* event for another day with corporate buyers and purchasing managers!). It really depends on how far upstream you take your value stream.

In the office, dwell time is important between steps as the longer information such as orders sit, the longer the order-to-cash cycle becomes.

The type of information gathered to develop a VSM also varies somewhat from an office situation versus the DC (see Fig. 11.2). The reality of mapping in a supply chain and logistics management environment is that it is usually a combination of office and warehouse (including transportation) value stream.

The data collection form shown in Fig. 11.2 gives you a good idea of the type of information required to create a VSM. Once the data is collected and displayed on a current state VSM, along with information flows, the team can get a good idea of where improvement opportunities exist.

The actual layout of the current state VSM can be seen in Fig. 11.1. The value stream is mapped from customer to supplier as previously mentioned. Data boxes are completed with information gathered for each process or activity in the middle of the map (with either inventory or dwell time between the activities), information flow shown on top, and individual and system-wide lead times (value- and non-value-added) at the bottom.

As the team discusses and reaches consensus that the current state map is accurate, they start to identify areas for improvement with the goal of reducing non-value-added activities. These opportunities are highlighted by "*kaizen* bursts" (shown previously in Fig. 11.1).

**Supply Chain and Logistics Management VSM
Data Collection Sheet**

- Process time
- Available time
- Setup time
- Lead time/turnaround time
- Typical batch size
- % Complete and accurate information (% C&A)
- Rework/revisions (e.g., design changes)
- Number of people involved
- Reliability (e.g., system downtime)
- "Inventory" – queues of information (e.g., electronic, paper) and/or physical inventory (raw, WIP or finished goods)

Figure 11.2 Supply chain and logistics management VSM data collection sheet.

In our example of the current state map for a purchasing process, the team found many opportunities to reduce non-value-added steps, such as a lengthy quotation and approval processes, which could potentially be improved upon.

Future State Value Stream Map

The future state map attempts to create a flexible, agile system that quickly adapts to ever-changing customer needs, eliminates waste, minimizes handoffs and "silos," and triggers resources only when needed (see example in Fig. 11.3).

As mentioned previously, the team should have gathered some thoughts on improvement during the information-gathering phase of the current state map. When starting to actually "put pen to paper" (or on a whiteboard) for the future state map, they should ask some questions such as:

▲ What does the customer really want or need?

This should include determining what "service level" the customer needs, response or turnaround times, required quality level of the output, expected demand rate and variation, and resources required to meet demand rate.

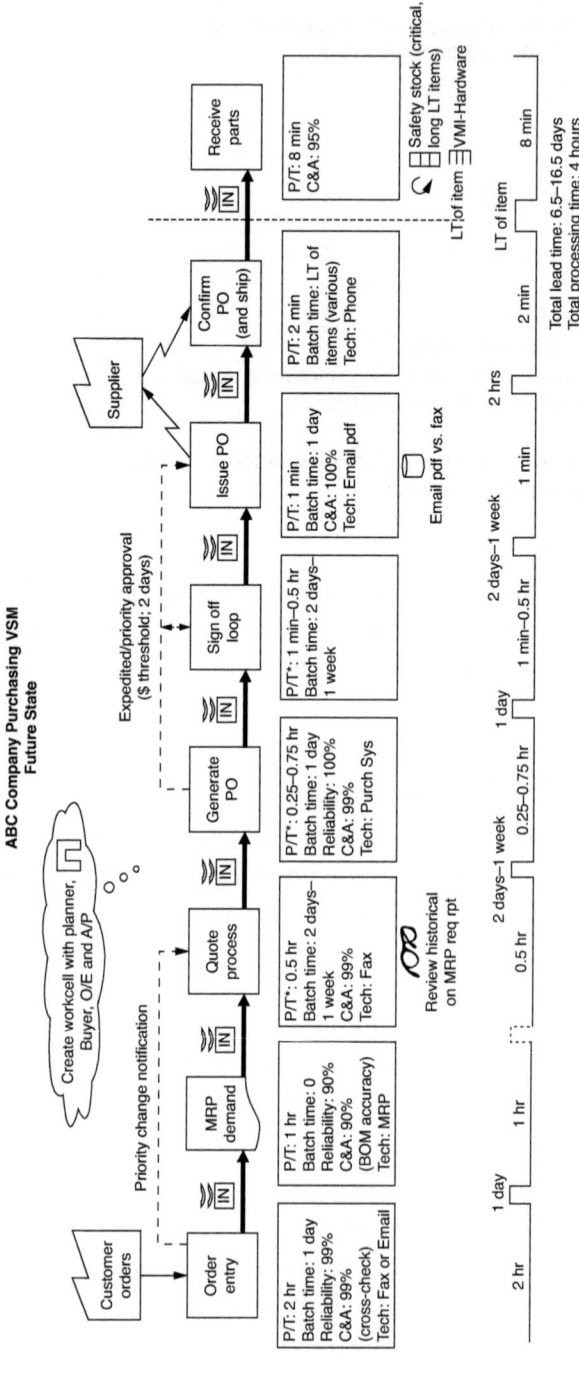

Figure 11.3 Future state value stream map example.

▲ How often will we check our performance to customer needs?

> This helps to define the frequency that the system will be reviewed to verify it is satisfying customer needs and the targeted cycle time or service level and how we will check the progress.

▲ Which steps create value, and which are waste?

> We should challenge every step and determine what is really needed by the customer, what can be done differently (or not at all). We also need to know if existing controls and administrative guidelines are appropriate and what knowledge and skills are truly required to perform the step(s).

▲ How can we flow work with fewer interruptions?

> We need to understand if we have continuous flow in areas such as customer service (e.g., from order to invoice), order processing (e.g., design-to-order), and warehouse and distribution for example. Identify bottlenecks in the process.

▲ How do we control work between activities?

> It is always useful to ask operators how they know what to do next when finished with a batch of work.

▲ Can we better balance workload and/or different activities and how will work be prioritized?

> We need to know if the "mix" (i.e., order/product/service type) impacts the ability of the system to flow, or impacts the responsiveness of particular steps in any way, and does the "volume" (i.e., demand variation) impact the system.

▲ Finally, what process improvements will be necessary and which are most important?

The answers to these questions will drive the future state map and implementation plan. This is also where the team can consider applying many of the basic and advanced Lean concepts we discussed earlier in the book such as layout, visual workplace, setup reduction, JIT, pull/*kanban*, work cells, etc.

Where to Look

Overall, the focus should be on eliminating waste in the supply chain, while trying to improve continuous flow. There should be an emphasis on shifting from a push to demand-pull process based upon customer demand (using concepts such as takt time, load leveling, line balancing, and one-piece flow) to minimize inventory. You should target more visibility into both the demand and supply chain to help manage service and costs.

As purchasing can account for 50 percent or more of the total costs, Lean tools can be implemented such as visual material management (e.g., *kanbans*, simple bin replenishment), vendor-managed inventory (VMI) systems, and supplier reviews/assessments to see how Lean your suppliers really are. There should be an emphasis on partnering, collaborating, and long-term alliances with sharing of cost and technical data and, in some cases, risk. All of which will require mutual trust.

In warehousing, waste can be found throughout, including: defective products which create returns; overproduction or overshipment of products; excess (or inaccurate) inventories, which require additional space and reduce warehousing efficiency; excess motion and handling; inefficiencies and unnecessary processing steps, excess transportation steps and distances; waiting for parts, materials, and information; and inefficient (and manual) information processes.

The final future state map should be reviewed with the team, then presented to upper management for their approval and support.

In our example of the future state map for a purchasing function (previously shown Fig. 11.3), the team determined it would be an improvement to create separate dollar-value "thresholds" for purchase orders, so as to improve the approval process time for smaller purchase orders. Among other thoughts, they also came up with a longer-term idea to create work cells that would speed up the entire process by combining the order entry, planner, buyer, and accounts payable roles into one process and improve flow.

Implementation Plan

The final step (and really just the beginning) is to prioritize and plan for improvements identified in your future state map to be implemented in the next 3 to 6 months. There is usually some "low-hanging fruit" that can be done quickly and easily. The harder improvements, which may take some capital improvements and more time, need to be prioritized and planned for (see Fig. 11.4).

Yearly Value Stream Plan

Signatures

Date:				Plant Mgr	Union	Eng	Maint
Facility manager:							
VS manager:							

Product family business objective	VS future state loop	VS objective	Goal (measurable)	Monthly schedule												Person in charge	Related individuals and Depts	Review schedule	
				1	2	3	4	5	6	7	8	9	10	11	12			Reviewer	Date
	1																		
	2																		
	3																		

Product Family:

Figure 11.4 Value stream plan.

It is a good idea to tie the future state map to the overall business objectives as mentioned previously.

It is also sometimes best to break a future state map into easier-to-digest and easier-to-deliver *loops*. For example, you may create separate *customer, warehouse,* and *supplier* loops.

To ensure success, you will need leadership and support from upper management, a company business strategy that includes Lean, development of a culture that emphasizes a continuous improvement mentality, and ways to ensure that Lean becomes part of everyone's job description and is discussed on a regular basis.

Additionally, it is always a good idea to benchmark key metrics within your industry and outside your industry and to always question the status quo.

Lean Teams

Figure 11.5 Team work.

In many ways, the Lean journey by itself is a form of team building, especially when using a simulation training game early on in the process to help trainees transfer the general concepts and tools into hands-on learning (Fig. 11.5).

In any case, much more progress can be made and maintained if Lean is part of everyone's job, rather than just a few. While it is important to have a Lean structure, with Lean champions, Lean coordinators, and even in some cases, Lean subject matter experts (in areas such as VSM, 5S, etc.), it is more important and effective to have everyone involved from top to bottom. Companies where Lean becomes part of the culture tend to get greater, longer-lasting results. While those who just look at it as the fad of the month may get some good results, but usually go back to doing things the way they were done before the training. What you really want to do is to develop a culture that consistently defines and solves problems utilizing Lean tools.

There are many reasons to work in teams, and in fact, it is pretty much the norm these days, including everything from new product teams to quality teams and beyond. Teams are a great way to share ideas, and they create a support system for the members. They also use the skill set of all of the members as the sum of ideas is usually better than an individual person's ideas. Besides, it is more fun to work in teams, and as a result, implementation is usually easier.

When creating teams and a teamwork culture it should be in a risk-free environment but should still be disciplined in terms of the process and rules. Trust is a critical factor, as well as selecting the right people for the team. I always prefer a mix of people from a variety of areas (e.g., production, supply chain, logistics, engineering, sales, etc.) to gain a common understanding and to avoid silo kind of thinking. I also prefer to take volunteers first.

Team Charter

It is always good to formalize things with a *team charter* (Fig. 11.6) so that everyone knows their roles and responsibilities.

The Team Makeup

You should make sure that you have a fairly high-level sponsor/advisor for the team to help break through any roadblocks along the journey, as well as being able to arrange for support for the team during the event(s).

The team leader is kind of like the quarterback who determines session objectives, the process to be followed and the agenda, and should meet with the facilitator to review session objectives and process.

The traits to look for in a leader include any previous success as a leader, some knowledge of Lean Enterprise (preferably hands-on experience), and someone who is comfortable working in the targeted area(s).

The consultant/trainer should be more of a facilitator. I typically deliver Lean concepts and application training but usually end up doing

Team Charter

Date:	_____
Target area:	_____
Target area supervisor/Manager:	_____
Mission/purpose:	_____
	_____
Team name:	_____
Champion/Facilitator:	_____
Leader:	_____
Team members and roles:	
Recorder:	_____
Time keeper:	_____
Members:	_____
	_____
Scope of project:	_____
Critical success factors (including metrics):	_____
	_____

Figure 11.6 Team charter form.

consulting as well to help the steer the team in the right direction in general, as well as help to come up with and organize specific improvement ideas.

The actual team members should be a blend of people from inside and outside of the area, as mentioned previously, and they should understand the target area (may work in it or can learn about it) and be open to thinking about doing things differently.

When it comes to meetings and their rules, it is important to set them early and try to adhere to them as much as possible. They should include basic things such as showing up and starting on time, being prepared, listening attentively, participating, etc., as well as individuals assuming responsibilities and supporting any group decisions.

It can be very useful to have a place on the whiteboard (or large piece of paper) to use as a *parking lot* for good ideas that may be off-topic at the moment, but have longer-term potential later.

Kaizen Events

Kaizen is loosely translated (or commonly known) to mean continuous improvement. So a *kaizen* event is a team-based continuous improvement project. The goal of a *kaizen* event is process improvement through the elimination of waste at all levels of the process. Typically, prior to the actual event, there is a period of training in general Lean concepts and applications, as well as additional training in specific tools, such as setup reduction, work cells, etc.

As mentioned earlier in the book, the first *kaizen* event is usually either 5S-workplace organization or implementing the results of a VSM project. However, you can also create a *kaizen* event based upon feedback from your team members after some basic training.

For example, after some basic Lean training, the general manager at a client facility that functioned as both a private DC for a cosmetic contract manufacturer, as well as a third-party logistics services (3PL) facility had his team do some brainstorming to come up with ideas for *kaizen* events. We handed out a form similar to the one shown in Fig. 11.7 and asked the individuals to come up with improvement ideas.

Cost Reduction Kaizen Implementation

Department:_____ Process for Kaizen:_____ Kaizen #:_____

Cost Center:_____ Date:_____

Approvals: Lean champion:_____ Maint:_____ Controller:_____ GM:_____

(1) Current situation	(3) Solution activity

(2) Analysis	(4) Cost reduction	(total savings:)
	Current	Proposed

Figure 11.7 Cost reduction *kaizen* implementation form.

The team came up with many good ideas including one idea for improving the cycle-counting process. Their current cycle-counting process was done manually, even though the DC had radio frequency (RF) scanning capabilities for receiving, picking, etc. Time studies showed that they were wasting an average of 108 minutes doing data entry per day and 54 minutes in wasted travel time to the office. The solution was to use RF devices for cycle counting, which would not only significantly reduce these wastes but also cut down on data entry errors.

In general, a *kaizen* event is appropriate in a number of situations including when there is a need for an urgent solution, competitive issues, customer service or cost issues, and bottlenecks.

A major difference between *kaizen* events in America and Japan is that in the United States there seems to be more of a rush to get results, versus the slow but sure way of continuously improving a process in Japan. That may be one reason that many Lean initiatives fail in the United States. They may make short-term gains, but don't stick with it for the long haul as Japanese companies do.

Kaizen Event Management

When planning for a *kaizen* event, there are some general steps that are good to follow:

1. Prepare for the event by selecting the targeted area and team.
2. Define scope and goals of event with team members.
3. Train the team in various Lean concepts and relevant applications.
4. Walk the *kaizen* area with team members.
5. Collect data on the *kaizen* event area (varies depending on whether it is for a VSM, 5S, or other specific tools to improve the process by removing waste.
6. Brainstorm ideas as a group.
7. Prioritize the top ideas in terms of value to the customer.
8. Form subteams to implement the ideas.
9. Keep track of progress and check results.
10. Develop/review/update employee job instructions where needed (adding standardized work where ever possible).
11. Develop an action plan for remaining ideas.
12. Regularly report plan, progress, and results to management.

13. Recognize the team and communicate results to entire organization.

14. Follow up on open action items.

15. Measure area improvement versus goals and objectives.

16. Disband team when *kaizen* is finished.

From an upper-management perspective, whether from a steering committee or Lean champion standpoint, besides each *kaizen* event being organized similar to the list above, it is key that there is some kind of *kaizen board* to know what the *kaizen* schedules and progress are to keep track. A *kaizen* board is an ideal tool to make the events visible and can be a simple cork- or whiteboard. They are useful for controlling progress, sharing information, and keeping motivation going.

A great first step after some general overview Lean training, a VSM is a valuable tool for continuous improvement. It serves as a "road map" to the future and is a true foundation for the entire Lean journey.

CHAPTER 12

Lean and Technology: Why Can't We All Just Get Along?

Historically, Lean has been viewed as a "pen and pencil" approach, one that is very visual and needs little help in the way of technology. However, if you think of technology "enabling" a process, then it can be of great help enabling a leaner process, especially in the supply chain and logistics management function, which is all about communications, collaboration, and visibility.

Lean and Technology: Background

A 2006 study by the Aberdeen Group entitled "The Lean Supply Chain Report—Lean Concepts Transcend Manufacturing through the Supply Chain," found that "as Lean moves from the plant floor to the supply chain, it becomes more difficult to orchestrate activities without automation. Although the Lean early adopters were not proponents of technology, circumstances have changed. The majority of manufacturers rely on a combination of corporate ERP and semi-automated Lean processes to support their business operating models." [www.aberdeen.com, 2006]

The study goes on to discuss the fact that less than 50 percent of Lean tools such as VSM, supply integration and planning (e.g., *kanbans*) and scheduling are "automated" using ERP, SCM, and "homegrown" systems with the remainder being more of a pen-and-pencil type of solution (this includes the use of spreadsheets).

However, the Aberdeen research noticed "...a consistent trend for IT solutions with the greatest impact on Lean strategies as customer facing: 'integrated manufacturing and logistics solution' and 'integrated order configuration/management and manufacturing solution.'" [www.aberdeen.com, 2006]

This study indicates that while there has been an increased use of technology in Lean programs, there is still room for continued growth.

According to a more recent Aberdeen Group report from 2009, entitled "Lean Manufacturing: Five Tips for Reducing Waste in the Supply Chain," organizations that have applied technology to Lean manufacturing today are improving speed, efficiency, and profitability.

Best-in-Class Use of Technology with Lean

In their survey of 117 companies, Aberdeen found that the "best in class" companies are using Lean principles and software solutions "as a long-term strategy for improving people, processes and business results." [www.aberdeen.com, 2009]

Aberdeen concluded that these measures, along with supermarket sizing and order management integration (which gives added visibility into manufacturing constraints taking customer orders) are among the "Lean automation" tools being used by manufacturing firms today.

As far as top Lean-enabling technologies go, the study found that 63 percent of "best in class" manufacturers (which make up the top 20 percent of companies) have enabled Lean manufacturing practices through demand planning and forecasting systems to improve production planning and scheduling, while 43 percent have used manufacturing execution systems (MES) and 42 percent have used advanced planning and scheduling (APS) systems. [www.aberdeen.com, 2009]

It should be noted that an MES is a control system for managing and monitoring WIP on a factory floor and an APS is a system in which raw materials and production capacity are optimally allocated to meet demand. The outputs of an APS are production plans at different levels of detail.

For purposes of this book, it works best to break the Lean SCM technology discussion into two chapters: This one is focused more on your *internal* SCM operations, and the next chapter is focused "beyond the four walls" of your operation. There is some overlap as you probably already realize, but we will go with it for now.

There is a lot of technology available today so it is difficult to go through all of it. We will at least try to cover the most common technology elements and how they can enable a Lean supply chain.

Enterprise Resource Planning (ERP) Systems

Of course the most common technology these days are ERP systems. ERP systems are really an integration of all business processes of an organization in one common database, typically in "real-time." They help to coordinate decision making all along the supply chain from customer to supplier. The (mostly) internal functions that ERP manages typically include: materials requirements planning (MRP), finance, human resources, SCM, and customer relationship management (CRM).

ERP systems can be very costly and need much customization, but, in terms of waste reduction potential, they can reduce transactional costs and increase the speed and accuracy of information. They can also enable JIT systems, another key Lean element.

ERP systems can also be integrated with other internal systems, such as warehouse management systems (WMS), forecasting, distribution requirement planning (DRP), and quality systems (in some cases, these modules may be included/available in an ERP system).

An IFS software white paper from 2009 by Jakob Bjorklund entitled "10 Ways to Use ERP to Lean the Manufacturing Supply Chain" pointed out some ideas that are very relevant to this discussion. [Bjorklund, 2009] Besides using various methods and tools to "Lean" out your supply chain processes as we've pointed out earlier in the book, the white paper points out some technology tools to look for in an ERP system.

While there isn't really one "silver bullet" to have in an ERP system, in terms of helping you on the Lean journey, there are some specific things to look for. Of course the system should be integrated company-wide, which is pretty basic to ERP systems these days.

It is critical it is to have an integrated quality management system to record and analyze quality performance, so you can "do it right the first time," as they say.

Demand forecasting, which we will discuss in more detail later in this chapter, is important for not only reducing waste by having more accurate forecasts to drive production, purchasing, and deployment of inventory, but it is an important communication and collaboration tool. So it is important that you consider a "best in breed" type of forecasting system, which may not be the case with what may come with an ERP system. In addition, if it is an external add-on system, make sure it is integrated with your ERP system. [Bjorklund, 2009]

The IFS software white paper also points out that:

▲ It is important that your ERP system "supports multiple modes, including make-to-order (MTO), engineer to order (ETO), configure to order (CTO), and others." This helps you to consider producing in batches when necessary (e.g., make-to-stock or MTS) or when the opportunity presents itself, in a more customized, small lot manufacturing mode.

▲ The application should be capable of being "demand-pull" based, minimizing raw, WIP, and finished goods inventory and maximizing customer service using tools such as distribution requirement planning (DRP) and quick response (QR)/collaborative planning, forecasting, and replenishment (CPFR), both of which will be discussed in Chap. 13.

▲ The ERP system should be capable of providing for multiple sites, not just your own, but distributors, wholesalers, and customers, in order to execute some of the technology just mentioned.

▲ Master data management is key to companies having multiple divisions that might also have multiple part numbers for the same items. Without this capability, visibility and integration can be severely impaired. [Bjorklund, 2009]

Demand Forecasting

As we all know, all forecasts are wrong. It is really a matter of having a collaborative process in place, which is a blend of "art and science" (or in other words, a mix of qualitative and quantitative methods) to make it as accurate as possible. As mentioned earlier, inventory is one of the eight wastes and covers up variability in a process, one of which is demand variability.

In Chap. 4, we discussed how it is critical to have a good planning process in place to minimize this variability and to minimize the bullwhip effect. Once this process is established, it is critical to have the right technology available to enable it.

According to an article in *SAScom Magazine*, entitled "The Future of Forecasting Software," "significant progress is taking place in the areas of automation, scalability, and the incorporation of structured judgment." [www.sas.com/news/sascom, 2006]

Trends Driving the Use of Technology to Reduce Waste

The article points out that there are trends in forecasting that are driving the need for increased use of technology including:

- ▲ The range of business forecasting problems is increasing—other areas of business require forecasts, such as warranty claims, returns, staffing, and maintenance, which can dramatically affect profits. So the more accurate and detailed the forecasts, the less waste there is in these and other processes.
- ▲ The scale of business forecasting problems is increasing—this has forced the use of more "automated" methods for forecasting and replenishment in places like retailing, for example, as a result of the proliferation of SKUs.
- ▲ It is crucial to distinguish the "high-value" forecasts for special attention while automating the "not-as-valuable" forecasts—this relates to previous discussion of using the Pareto method (ABC) to distinguish between your A or revenue/profit producing items and your slow-moving, less profitable C items. The A items require more time and effort in planning, whereas the C items, which are less important, can rely more on an automated forecast.
- ▲ There is a need to handle the "continuously evolving product"—as technology enables ever shortening product life cycles, there is less demand history available, and therefore there is a need for technology to intelligently look at the history of similar items to develop new forecasts (along with a tool that allows for collaboration).
- ▲ Quantity and quality of data will continue to increase—technology has made it easier to capture greater amounts of more detailed information using technology such as POS (point of sale), RF (radio frequency) bar code scanning, RFID (radio frequency identification) and the Internet, creating huge "data warehouses" of information. Therefore, there is a need for more automated forecasting systems to compile and interpret this information.
- ▲ Use of structured judgment enhances collaboration—with techniques such as CPFR and VMI, there are more sources of data input from more participants, including remote sales, customers, and suppliers. [www.sas.com/news/sascom, 2006]

The article concludes that with the use of better forecasting tools and methods, forecast error (a source of waste) will be reduced, but demand will become more erratic due to SKU proliferation, shorter life cycles, etc. This takes us back to previous comments where we pointed out that the optimal way to handle this, besides the use of technology to improve forecast accuracy, is through more flexible, Lean Manufacturing and supply chain processes.

Distribution Requirement Planning (DRP)

DRP, similar to MRP, is time-phased planning, except in this case it is used to manage finished goods in your distribution network. DRP is ideal for businesses with complex distribution networks, but also effective in managing inventory in simpler ones. DRP enables businesses to go from a demand push, to a pull environment, thereby reducing waste in the form of inventory, transportation, and warehouse costs, as well as time by triggering replenishment based upon the pull of the customer order.

DRP and Demand Pull

As shown in Fig. 12.1, typical DRP software reviews on-hand inventory levels, which are netted against open customer orders, forecasts, work/purchase orders, and pending transfers and determines if replenishments are needed (when the net inventory drops below a targeted safety stock or safety time level).

By pulling inventory through in this way, inventory is minimized (and customer service levels maximized) and product is delivered more "just in time" instead of "just in case." DRP also helps to improve supply chain visibility so that companies can plan rather than react.

For example, imagine an item or SKU that has very seasonal demand such as snow shovels. If you use a tool such as DRP, it will pick up seasonal demand (planned, actual, or some kind of combination) in advance based upon the lead time for that item. The lead time can represent the transit time to the distribution center or the manufacturing or procurement lead time, or both as DRP is "hierarchical" (i.e., multiple levels of demand). DRP allows you to react quicker to changes in demand at the retail level, as well as to improve your production and distribution planning process through more accurate, pull-based replenishment forecasts.

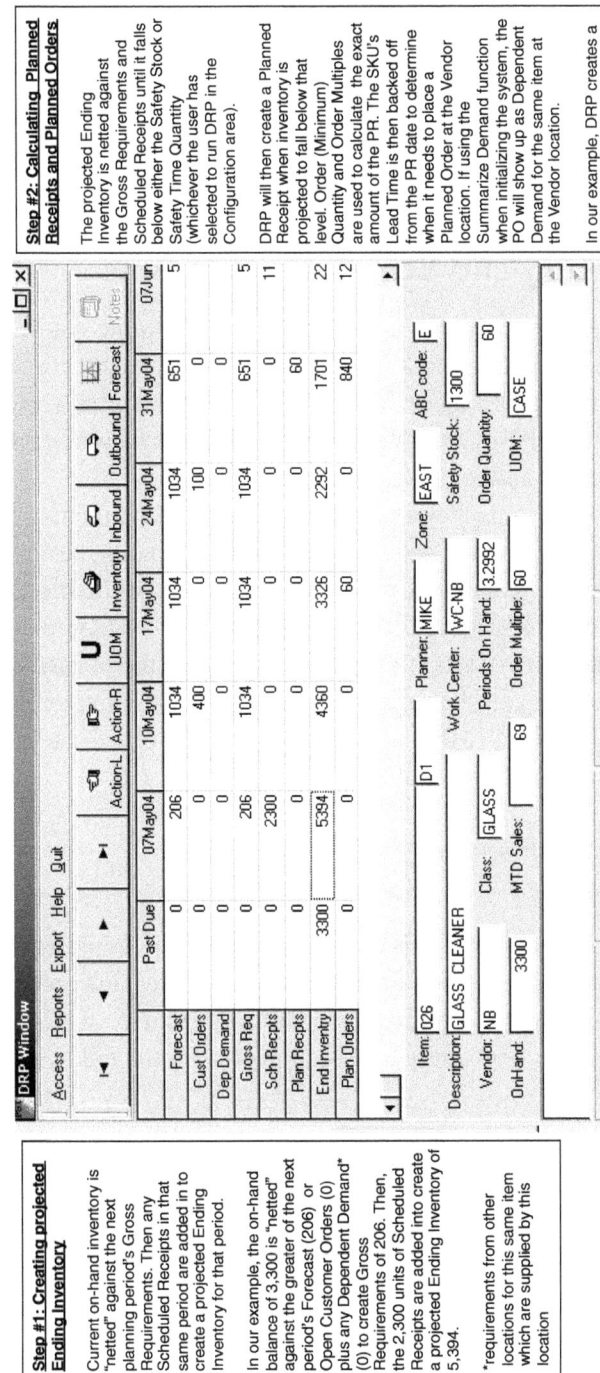

Step #1: Creating projected Ending Inventory

Current on-hand inventory is "netted" against the next planning period's Gross Requirements. Then any Scheduled Receipts in that same period are added in to create a projected Ending Inventory for that period.

In our example, the on-hand balance of 3,300 is "netted" against the greater of the next period's Forecast (206) or Open Customer Orders (0) plus any Dependent Demand* (0) to create Gross Requirements of 206. Then, the 2,300 units of Scheduled Receipts are added into create a projected Ending Inventory of 5,394.

*requirements from other locations for this same item which are supplied by this location

Step #2: Calculating Planned Receipts and Planned Orders

The projected Ending Inventory is netted against the Gross Requirements and Scheduled Receipts until it falls below either the Safety Stock or Safety Time Quantity (whichever the user has selected to run DRP in the Configuration area).

DRP will then create a Planned Receipt when inventory is projected to fall below that level, Order (Minimum) Quantity and Order Multiples are used to calculate the exact amount of the PR. The SKU's Lead Time is then backed off from the PR date to determine when it needs to place a Planned Order at the Vendor location. If using the Summarize Demand function when initializing the system, the PO will show up as Dependent Demand for the same item at the Vendor location.

In our example, DRP creates a PR for 60 units to maintain a Safety Time of 3 periods of future Gross Requirements (1,701). The Lead Time of 2 planning periods is backed off to determine when the replenishment needs to ship from the Vendor location

Figure 12.1 Sample DRP screen and description.

Source: PSI Planner for Windows™, Copyright 1998–2011.

This type of tool, along with a forecasting system, can also be extended to include customer locations (e.g., customer distribution center and retail locations) to reduce waste even further by improved forecasts and visibility downstream in your demand chain. We will discuss in more detail in the next chapter when exploring CPFR systems.

Advanced Planning and Scheduling Systems

We will just speak briefly about *advanced planning and scheduling systems* (APS) since they actually fall more under the heading of manufacturing as they are typically used for planning or scheduling production. However, when integrated in an S&OP process, they can have a direct and powerful impact on the supply chain and logistics process.

APS systems can include an array of software used in the manufacturing management process where raw materials and production capacity are optimally allocated to meet demand. They are typically used in situations where there are complex trade-offs, sequencing optimization, and competing priorities, all of which are constantly changing. The systems can assist the planning process starting at the capacity or aggregate planning level, master production scheduling (MPS), and even detailed finite capacity scheduling (FCS) or short-term scheduling levels.

Benefits of Advanced Planning and Scheduling Systems

A robust S&OP process can help to drive a Lean process from a management perspective, but in order to do so, it takes the support of some pretty heavy-duty detailed, complex planning and scheduling systems.

A good APS system can contribute to a Lean process in the following ways:

▲ Reduce cycle time as a result of decreased waiting time and WIP inventory between manufacturing operations.
▲ Reduce raw material carrying costs.
▲ Reduce finished goods material storage by using storage space as a critical resource.
▲ Improve on-time delivery by showing precisely when jobs will finish.
▲ Improve accuracy of order promise dates.

▲ Increase utilization of key resources.
▲ Streamline and standardize the schedule process by storing all of the rules in one system.
▲ Centralize information to improve communications between functions.

Warehouse Management System (WMS) Software and Radio Frequency Identification (RFID)

WMS is a software tool for managing inventory within the four walls of a distribution center. It typically is integrated with a company's ERP system, which can be updated in batch or in some cases "real time." WMS software typically covers planning and movement of inventory from receipt to shipment (i.e., receiving, put away, picking, staging, and shipping). The functionality typically found in a WMS includes: inventory control, storage location management, quality control interfacing, order picking, automated inventory replenishment, receiving, shipping, operator productivity, and report generation.

Data can be manually entered into a WMS, or more optimally electronically via RF devices (or more advanced RFID).

Warehouse Management Systems in Waste Reduction

WMS can reduce waste in many ways. The reduction of errors, especially when using bar code scanning equipment, improves the accuracy of inventory, including the accurate receipt, shipment, and location of items in the DC. It directs activities from incoming material processing to manufacturing and through order fulfillment, making sure that the correct materials are getting to the right place, at the right time, in the proper quantities to satisfy customer orders. Labor, the productivity of which is critical to controlling cost in a DC (e.g., cases/hour), is system-directed in an optimal manner and based upon general or customer-specific business rules. So features such as directed put away, wave/batch picking, velocity slotting, and pick to light help to improve both accuracy and productivity.

Radio frequency identification (RFID) is a lot more than just a small improvement from bar-code technology. Bar codes gave us a status report at a certain point in time and are an automated form of data entry, whereas

RFID, in the long term, will potentially provide complete and continuous visibility throughout manufacture, shipping, warehousing, and sales of a pallet, a case, or a single item. The information will be received much faster than with just bar codes and offer much more information about the item (and use less labor!). As the RFID chips become more sophisticated, less expensive and the data more standardized, there will be even more efficiencies.

Transportation Management Systems (TMS)

In its simplest terms, a TMS is a software system that helps to manage a company's transportation operations. They can help to select the most cost-effective mode and routes that match inventory and delivery objectives.

The functions that this type of system can offer include:

▲ Planning and optimization of transportation routes
▲ Transportation mode and carrier selection
▲ Management of carriers
▲ Real-time vehicles tracking
▲ Service quality control
▲ Vehicle load and delivery route optimization
▲ Transport costs (including rating, pre- and post-audit)
▲ Shipment batching of orders
▲ General cost and performance control measuring key performance indicators (KPIs) with reports and statistics

Benefits of Transportation Management Systems

The benefits of TMS in conjunction with Lean can include decreases in transportation costs, inventory levels, and carrying costs, as well as increases in asset utilization, supplier fill rates, and customer service.

According to a 2006 Aberdeen Group survey "The Transportation Benchmark—The New Spotlight on Transportation Management and How Best in Class Companies Are Responding," "transportation management is moving out of the shadows and into a strategic role in driving supply chain excellence. In recognition of this, most companies are actively reevaluating their transportation management processes, organizational structure, and technology." [www.aberdeen.com, 2006]

The survey found that the most important actions were really waste reduction—focused as they included getting accurate online status and cost information, gaining visibility into shipments and orders, and collaborating with carriers, suppliers, and customers to create more economical transportation processes.

TMS software can increase efficiency across the company through better shared arrival status (inbound and outbound), and visibility for more employees and customers, for example. More timely, accurate information leads to less uncertainty and more accuracy financially, as well as by not only reducing transportation costs, but also by reducing inventory holding costs and having more predictable lead times.

The Aberdeen Group survey also goes on to say that on-demand, Web-based solutions can be implemented faster and cheaper and gives quick and easy access to real-time information to all shareholders (internal and external). This reduces waste in many ways, including substituting real-time information for excess inventory and reduces operational costs as well through better efficiency, but there are always advantages and disadvantages to going with this type of solution so a lot of time should be spent researching it before jumping in.

It is the opinion of Kelly Thomas, senior vice president of manufacturing at JDA Software Group, Inc., that no matter which specific technology you go with, in general, it "will play a critical role in the ability of companies to adapt to constantly changing business conditions. Companies that fail to leverage technology—or to leverage it properly in the context of people and process—will fail to maintain their competitive edge. Dramatic shifts are occurring in many industries—retail, high-tech electronics, high-tech semiconductor, automotive and metals, to name a few. Companies that fail to recognize these shifts will get caught behind the change curve. Many of those that fail to recognize the shift will be left behind. The current shift requires demand-driven, lean SCM." [Thomas, 2011]

CHAPTER 13

Beyond the Four Walls: I Can See Clearly Now

In Chap. 12, we discussed various technologies that can be used to reduce internal waste in a business. However, if we do not look at the entire supply chain network, we run the risk of passing off our inefficiencies to customers and suppliers, resulting in them becoming less efficient. For example, by just reducing our lot size on orders from a supplier without other communications, the supplier does not really get the necessary visibility into our schedule to determine the most efficient action to take when producing the material for our order.

In many cases, we must also interface with partners to improve overall supply chain efficiency and flexibility. As the term supply "chain" indicates (actually more like a "web"), we are all a network of entities that affect each other's performance as well as impact the final customer. So the more visibility we have into each other's operations (our supply *and* demand chain), the more waste we can eliminate.

This is easier said than done, especially if we do not have the proper tools to enable this visibility and communicate it properly. That is where technology comes in. Some of it, like electronic data interchange (EDI), has been around for a long time, whereas tools such as *software as a service* (SaaS—the "on-demand" rental of software that is installed at a remote, dedicated, secure server) are relatively new. Thanks to the almost exponential improvements in technological capabilities and capacity combined with decreasing costs for such technology, even small- to medium-sized companies can now take advantage (much of this can be attributed to *Moore's law*, which states that the number of transistors on a chip will double about every 2 years).

Electronic Data Interchange (EDI)

The first tool to discuss is EDI, which enables companies to exchange documents used in intercompany processes, such as purchase orders and invoices, in a structured electronic format easily processed by computers. This helps to automate and streamline business by eliminating or simplifying clerical tasks, speeding information transfer, reducing data errors, and eliminating business processes.

EDI standards have been developed over the years, independent of software and technology. There are standards for various activities. For example, in the United States, ANSI ASCI X12 is the standard used with specific formats for purchase orders (assigned an identifying number of 850) and invoices (810).

EDI has been successfully used in some specific industries, such as retail, and in some larger companies since the 1980s, but it has not been widely adopted by many small- to medium-sized enterprises. The primary barriers to widespread acceptance of EDI have been the costs of implementation and of communication, which are frequently accomplished using value-added networks (VANs). These costs are generally too high for companies (or "trading partners" as they are called) that do not conduct large numbers of EDI transactions.

Web-Based EDI

In recent years, Web-based Internet EDI has emerged, providing valued-added functions traditionally provided by EDI over VANs. However, most EDI transactions are still handled by VANs. Web-based EDI allows a company to work with its suppliers without having to implement a complex EDI infrastructure. Simply put, Web-based EDI lets small- to medium-sized businesses receive, create, send, and manage electronic documents using just a Web browser.

In fact, according to Robert L. Sheier in the "Internet EDI Grows Up," Walmart announced that "it would do electronic data interchange (EDI) with suppliers over the web instead of using value-added networks (VAN), it was a signal that EDI over the web is ready for heavy-duty corporate use after years of development….it requires customers to do some of the work the VAN used to do, such as making sure the proper person is notified if a purchase order or invoice doesn't get through. However, if customers can cost-effectively become their own VANs and choose the right web EDI tools, the savings can be compelling." [Sheier, 2003]

A surprising fact pointed out in the Laudon and Traver text *E-Commerce* (Prentice-Hall, 2009) is that business-to-business e-commerce, which is primarily driven by EDI, is the largest form of e-commerce with over $3.8 trillion in transactions in the United States in 2008. This compares with the relatively small, but rapidly growing business-to-consumer e-commerce (the one we arere more familiar with) at only $255 billion in 2008. [Laudon and Traver, 2009]

EDI has the obvious waste reduction benefit of fewer errors because of less people keying and rekeying information, as well as the element of reduced time to communicate and interpret the data, thereby shortening cycle times and ultimately the order-to-cash cycle.

It also reduces the technological bottleneck of different systems trying to "talk" to each other in different languages. EDI standards enable them to all talk in the same language leaving less room for interpretation and error.

EDI also enables quick response (QR), efficient consumer response (ECR) and collaborative planning, forecasting, and replenishment (CPFR).

E-Commerce

Beyond EDI, the ever-increasing use of the Internet for transacting business is helping to transform the supply chain. Thanks to e-commerce, the consumer now has the upper-hand, leading to more and more mass customization and true demand-pull. In addition to quality and cost, responsiveness and flexibility are paramount in this new world.

According to the book *Enterprise E-Commerce* by Peter Fingar et al., "an e-commerce platform allows an enterprise to extend supply chain automation to its suppliers' suppliers and its customers' customers, forming dynamic trading networks: end-to-end supply grids containing real-time business process facilities and shared data warehouses of information for decision support..." and that "multi-divisional companies typically operate multiple supply chain management systems to handle multiple plants and distribution channels... Value chains are being made into multiple-path, multiple-node value Webs. An extended SCM system can allow traditional, tightly linked systems to share information across channels and provide new opportunities for optimization across multiple, external supply chains." [Fingar et al, 2000]

E-Commerce and Small- to Medium-Sized Enterprises

E-commerce also positively affects small- to medium-sized enterprises (SMEs), putting them on more equal footing as the "big guys." Fingar et al. also point out that

> ...small to medium enterprises represent a whole new world of potential suppliers that can be tapped as a result of the Internet smashing the barriers of cost and complexity of traditional EDI-based systems. Supply, demand, and production planning and logistics can be optimized by extending automation opportunities to SME suppliers. Even the smallest SME will likely have access to a fax machine and a web browser. Because these simple touch points can be reached by the web, SCM business processes can be extended to virtually any SME, anywhere, anytime..." concluding that "...in the future, supply chains will function as a real-time business ecosystem. The richness and low cost of the Internet makes it possible to add new collaboration links with existing suppliers—and their suppliers—for forecasting, logistics, replenishment, bidding and ordering. A given supplier may participate in multiple supply chains, and integrating information from them can give the supplier a consolidated information base for planning and operations. Collaborations can be ongoing or ad hoc in response to market events and conditions" and that "...customer-facing applications come online they must be integrated with the extended supply chain management systems. Ultimately, as such applications go live, the results can be customer-centered supply chain management..." [Fingar et al., 2000]

So it is not hard to see that increased use of e-commerce, in general, can make your supply and demand chain more efficient, flexible, and responsive—that is, it helps to reduce waste.

QR, ECR, and CPFR

Business collaboration has increased significantly in the past 20 years with the advent of EDI, the Internet, and sophisticated supply chain planning (SCP) tools. Each of these tools can help companies to minimize waste in their processes as well as their customer and supplier's processes.

Efficient Consumer Response

Efficient consumer response (ECR) began in the grocery industry in the 1990s. It involves trade and industry groups working toward making the grocery sector more responsive to consumer demand, while trying to reduce costs from the supply chain.

I was involved in ECR in the early 1990s with Church & Dwight, Arm & Hammer Division and their clients, H.E. Butt, a Texas-based grocery chain and then Wakefern (also known as ShopRite), a New Jersey–based grocery chain. We used a combination of EDI and PC-based forecasting and replenishment software to import customer data such as POS (point of sale) at the retail level, and on-hand inventory, shipments, and open purchase orders at their distribution center locations. All of this was to ensure that our company's products were always available at the retail locations. Basically, we used this information to place H.E. Butt's and Wakefern's orders for them, saving that expense, while improving inventory turns in their DCs and on-shelf availability of our products on their shelves (all potential sources of waste).

Initially, this added cost to us to operate the ECR process, but it eventually helped us to improve our own sales forecasts and inventory location placement as the actual POS demand was fairly steady and predictable. If a supplier can manage to get a "critical mass" of volume on an ECR system (e.g., top-20 customers), and integrate it with their internal planning processes, they should be able to minimize the bullwhip effect by having smoother, more accurate forecasts.

Quick Response

Another form of collaboration, which was pretty much identical to our ECR program, quick response (QR) was implemented initially by the apparel industry in the 1980s. QR, a program developed by textile and apparel manufacturers and retailers in the 1980s used various strategies to reduce inventory levels, improve merchandise quality, increase worker productivity, increase inventory turnover, and reduce merchandise markdowns and inventory costs. QR gathers data about consumer preferences with a goal of integrating the information into production schedules. QR is driven by POS data. Using technology such as EDI, the data is communicated upstream to influence planning decisions as a response to what consumers demand. The net result of QR is that textile mills, apparel manufacturers, and retailers collaborate to response efficiently to consumer demand.

Again, while at Arm & Hammer, we started a QR program with Kmart. This one did not start as smoothly, as one of the results of this type of program—smaller shipments more often—was not anticipated (surprisingly) by Kmart. The net result was that they had a long line of less-than-truckload (LTL) carriers lined up at their DCs. Eventually, Kmart got smarter and narrowed the program down to their top vendors, thereby alleviating the problem to some degree.

However, in the end, the same, positive benefits resulted for both the manufacturer and retailer using QR as were found for ECR, such as:

▲ Improved forecasts ultimately resulting in better retail shelf presence (i.e., less stockouts)
▲ Higher inventory turns and order fill rates
▲ Lower ordering costs

ECR versus QR

According to a 2002 study at the MIT Center for Transportation and Logistics, entitled "The Value of CPFR," ECR is "focused on category management (enhancing the effectiveness of the demand creation and satisfaction process through better promotions, new product introductions and store assortment); product replenishment with high consumer service and low inventories; and the development of enabling technologies." [Sheffi, 2002]

QR "continually meet[s] changing requirements of a competitive market place, which promotes responsiveness to consumer demand, encourages business partnerships, makes effective use of resources and shortens the business cycle throughout the chain from raw materials to consumer." [Sheffi, 2002]

The study found that: "both of the ECR and QR initiatives were slowly adopted across the respective industries that spawned them. They did help change attitudes and create the realization that companies must look beyond their own boundaries to achieve high level of customer service and low costs. The collaborative aspect of these processes, however, were never implemented as originally envisioned on a large scale, mainly due to the cultural difficulties associated with collaborative management and the lack of scalable software." [Sheffi, 2002]

Collaborative Planning, Forecasting, and Replenishment

Collaborative planning, forecasting, and replenishment (CPFR) is a more detailed and comprehensive type of relationship than QR and ECR. In the 1990s, the Uniform Code Council (UCC) created a formal approach to the CPFR concept with a nine-step model that involves trading partners establishing a joint business plan after which a collaborative planning process develops a joint sales forecast, followed by a corresponding replenishment forecast.

Even CPFR has failed to scale with many manufacturers and retailers. They still seem to be focused on the larger customer/supplier relationships. There may be many reasons for this, the greatest of which is that there is somewhat of a disconnect between manufacturers and retailers. It is not easy to balance retail replenishment requirements with manufacturer's production and shipping requirements.

An AMR Research study in 2001 ("Beyond CPFR: Collaboration Comes of Age") showed some of the positive benefits of early CPFR adopters including, for retail:

- Store stock rate improvements of 2 to 8 percent
- Inventory reductions of 10 to 40 percent
- Higher sales of 5 to 20 percent
- Lower logistics costs of 3 to 4 percent

The same study found manufacturing benefits, such as:

- Inventory reductions of 10 to 40 percent
- Replenishment cycle improvements of 12 to 30 percent
- Sales increases of 2 to 10 percent
- Customer service improvements of 5 to 10 percent

There is no doubt that there is a lot more collaboration going on between manufacturers and retailers today, thanks, in part, to better, less expensive technology, but there is still a way to go to get to that critical mass where it really pays off for both parties.

Vendor-Managed Inventory

Another, similar, form of collaboration is vendor-managed inventory or (VMI), which focuses more on a more efficient, leaner form of parts and supply replenishment. A typical VMI process involves a supplier automatically replenishing their customer's parts or supplies. To do this, greater visibility is required on the part of the vendor, usually via a communication tool, such as EDI, or a more "low-tech" way such as regular visits to the customer's facility (this form of VMI has actually been around for over 50 years…think of the snack vendor who visits the local convenience store and fills the racks and leaves an invoice behind).

Based upon mutually agreed-upon replenishment points (*kanban* systems are great for this), the vendor automatically replenishes inventory at a customer's facility. This greater visibility into the customer's inventory and possibly production schedule usually leads to better inventory turns and less stockouts. On the part of the customer, while they are giving up some responsibility and control, they are also reducing costs by not having to check inventory and place orders, in addition to the previously mentioned benefit of improved inventory turns and less stockouts.

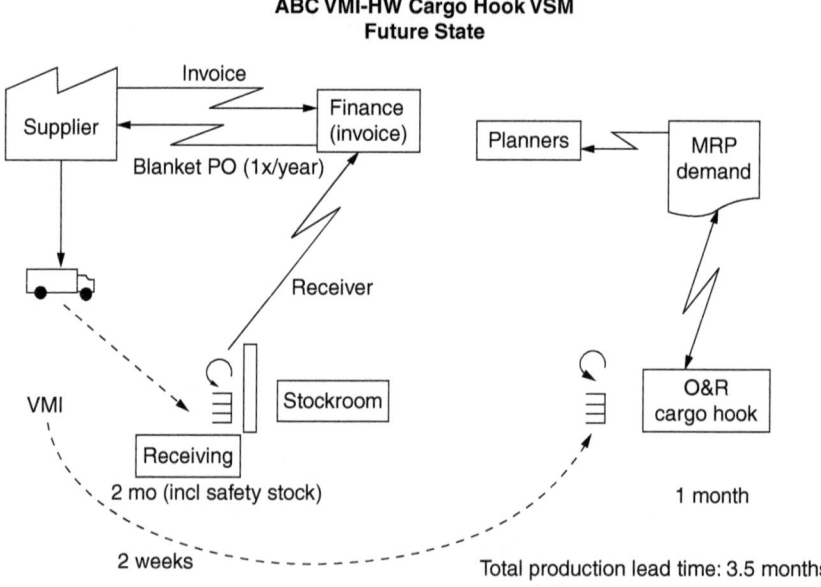

ABC VMI-HW Cargo Hook VSM
Future State

Figure 13.1 Parts VMI value stream map.

Value Stream Map for a VMI Program

Figure 13.1 shows an example of a VSM for a proposed parts VMI program. It would heavily involve the use of multiple *kanbans* with communication and collaboration as key components. When implementing a VMI program, it is best to work with suppliers who have had some experience in this area.

Other Potential Areas for Collaboration

There are other areas of potential collaboration requiring the enabling ability of technology, which were neatly summarized by Yossi in his 2002 paper "The Value of CPFR." They include:

▲ **Demand management**—going further than just forecasting production, planning, and replenishment; including collaborative merchandising, category management, promotional planning, and space management (at retail and DCs). Collaborative product design and new product introductions are already taking place in some companies.

▲ **Fulfillment**—extend the reach of CPFR into areas not touched, such as transportation carriers, forwarders, and public warehouses.

▲ **Real-time collaboration**—many of the current activities (CPFR, ECR, VMI, etc.) are more planning in nature. Many of the issues develop in "real time," which can even further enhance the value of collaboration with better visibility. [Sheffi, 2002]

Future Opportunities and Roadblocks

While collaboration "outside the four walls" with customers, suppliers, 3PLs, etc. may be the final frontier in terms of Lean, it is good to look at the challenges ahead.

A 2007 Supply Chain Management Review found that better flexibility in the supply chain was a result of strategic alignment, supplier integration, planning effectiveness, and relationship management technology. By "flexibility," they were referring to it in terms of order fulfillment lead time, supply chain response time, and production flexibility. [www.scmr.com, 2007]

Supply Chain Flexibility Traits

The study found the following traits within each competency that contribute to supply chain flexibility:

▲ **Strategic alignment**—clear supply chain (SC) goals and objectives driven by business strategy; business strategy exploits SC capabilities, and constraints and strategies communicated to all employees.

▲ **Supplier integration**—develop relationships to build on key supplier capabilities, exchanging operational information, synchronizing activities with suppliers and continuously exploring new working relationships.

▲ **Planning effectiveness**—formalized, disciplined processes addressing both long- and short-term planning, contingency and risk analysis with scenario evaluations, feedback loops addressing variances and vulnerability and continuity planning.

▲ **Relationship management technology**—CPFR, ECR, supplier performance, etc.

In other words, it is best to have a Lean process throughout your supply chain before using technology to enable it. But to be truly effective, you have to have all four traits to be successful, and technology can be of great assistance, especially in the areas mentioned in this chapter.

In the same survey, another competency found in Lean companies—better cost performance—indicated that both internal *and* customer integration were critical to reduced cost in the supply chain. The customer integration traits (which are best enabled by technology) found to contribute to this competency included:

▲ Relationships beyond sales transactions
▲ Planning for individual customer requirements
▲ Synchronizing activities with customers
▲ Continuously exploring new working relationships

It is readily apparent that the increased use of technology both internally and externally is the way forward to enable a leaner supply *and* demand chain.

CHAPTER 14

Metrics and Measurement: How Are We Doing?

According to the findings of the Bain & Company survey and report "Why Companies Flunk Supply Chain 101" by Myles Cook, more than 85 percent of senior executives say improving their supply chain performance is one of their top priorities, but fewer than 10 percent are adequately tracking that performance. Only 15 percent of the companies surveyed said they had *full* information on supply chain performance at their own companies, and only 7 percent go outside their four walls to track performance of supply chain activities at their vendors, logistics providers, distributors, and customers. [Cook, 2011]

Another Bain & Company survey of 300 global companies (source: www.bain.com) stated that "68 percent of managers think they have failed to optimize their supply chain savings. The ones who do—Walmart, Ford Motor, Dell Computer—all quantify performance indicators for their supply chains by setting targets that push them toward best-in-class status." [www.bain.com, 2011]

The feedback from Bain & Company is surprising to say the least. Now that we have got your attention, what is it we are supposed to look at?

Policies and Procedures

In general, the performance of a supply chain is the result of policies and procedures that drive various critical segments of the supply chain. The question is, "How can we design metrics to manage organizations recognizing that these organizations are components of complex and highly interconnected systems?" This question is rapidly gaining importance as supply chain managers face increased pressures on customer service and asset performance. Sony, for instance, is very aware of the fact that any inventory of its products at Best Buy and Walmart ultimately affects its profitability if it remains on the shelf for more than a few days. Sony has changed its delivery metric from "sell-in" to "sell-through." The difference is that the former metric allowed its sales department to chalk up a sale when the product was shipped to the customer (Best Buy, Walmart, etc.), whereas the latter metric chalks up a sale only when the product is sold and paid for. This is kind of like the "dock to dock" time measurement used in Lean Manufacturing. To give another example, Procter & Gamble uses its VMI process to routinely measure both its own inventory and the downstream inventory of its products.

Rationale for Considering Metrics

In a September 2007 *Industry Week* article "Seven Steps to Building a Lean Supply Chain," Mandyam M. Srinivasan pointed out that as a useful guideline when developing metrics it is worth asking whether a metric under consideration:

1. Helps sell more products, profitably
2. Helps reduce investments in resources
3. Helps reduce payments or expenses over the long term

If the answer to all these questions is no, then that metric should be questioned.

There are many supply chain metrics, some of which can indicate how Lean you are. We will discuss some of them now.

Relevant Lean Supply Chain and Logistics Metrics

The SCOR model (source: www.supply-chain.org) that we discussed earlier in the book also can be integrated with your supply chain metrics as they relate to Lean. SCOR has come up with five performance attributes, all of which can be related to various forms of waste. They are: delivery reliability, responsiveness, flexibility, cost, and asset management.

Delivery Reliability

Under the category of delivery reliability, we can look for waste in terms of shipping the correct product to the correct place and customer at the correct time. This also includes looking at whether or not we have shipped the product in perfect condition and packaging, in the correct quantity with the correct documentation. The resultant metrics measured would include:

- ▲ **Delivery performance**—Did it both ship and deliver to the client when they originally wanted it. Some companies adjust the delivery date based upon availability, change the date in their system, and measure performance based upon the new delivery/promised date. This results in an inaccurate view of delivery performance.
- ▲ **Order fill rate**—It is important to know if an entire customer order shipped complete. This metric is typically a lower percent performance than line item fill rate, which should also be measured.
- ▲ **Accurate order fulfillment (at various levels of detail)**—This is a quality measurement that looks at shipping errors, such as the wrong order or item(s) shipped to the customer (or the wrong quantity of requested items).

Perfect Order Measure

The culmination of this is the *perfect order measure*, which calculates the error-free rate of each stage of a purchase order. This measure should capture every step in the life of an order. It measures the errors per order line. For example, consider the following measurements:

- ▲ Order entry accuracy: 99 percent correct
- ▲ Warehouse pick accuracy: 99 percent
- ▲ Delivered on time: 95 percent
- ▲ Shipped without damage: 98 percent
- ▲ Invoiced correctly: 99 percent

Our perfect order measure in this case would be 90.3 percent (99 percent × 99 percent × 95 percent × 98 percent × 99 percent). This can be a challenging goal to meet when set at a high level, but it is a valuable form of measurement that points out the interrelationships between different aspects of your supply chain and gives a good idea as to how Lean your total supply chain really is.

Responsiveness

Responsiveness measurements relate to how quickly your supply chain and logistics function can deliver products to the customer. They can include measurements such as order fulfillment lead time, transit times, on-time delivery, and even overall cycle or dock-to-dock time (total time key material sits in a facility, which is a good measure of how Lean your organization is).

Flexibility

This is a measure of your supply chain's agility and response time when there are changes in the supply chain. As we know, there can be many unanticipated changes caused by economic, environmental, political, and other issues that make this something that can be used for a competitive edge.

Cost

It is, of course, important to manage your supply chain and logistics costs as that is a sign of potential waste. These measures would include cost of goods sold (COGS), total supply chain and logistics cost (in dollars and as a percent of revenue), transportation and distribution costs, warranty/returns, and a host of other individual costs.

Asset Management

These metrics look at how effectively a company manages assets to meet demand. This includes fixed assets and working capital. Metrics include order-to-cash cycle, inventory, and asset turns.

Balanced Scorecard

As there are literally hundreds of potential metrics to measure in the supply chain and logistics function, the use of the *balanced scorecard* approach can help to narrow it down.

A balanced scorecard is a tool that comes from the principles in the original Malcolm Baldrige Quality Award Criteria, stating that effective leaders take a *balanced* look at key results measures of an organization instead of relying too much on financial measures, which provide an historical look at organizational performance. So the basis for this tool is that business results are integrated and that management should not view one measure by itself without considering the relation to other results. A balanced scorecard looks at four different views of the business:

1. **Financial**—To succeed financially, how should we appear to our shareholders?
2. **Customer**—To achieve our vision, how should we appear to our customers?
3. **Internal business processes**—To satisfy our shareholders and customers, at what business processes must we excel?
4. **Learning and growth**—to achieve our vision, how will we sustain our ability to change and improve?

Objectives, measures, targets, and initiatives are developed for each of the identified perspectives to ensure success.

Finding the Right Metrics for Your Company

As competition increases and market forces continually change, supply chain performance management is a critical area for companies to help sustain and gain competitive advantage by enabling an agile, Lean, and efficient customer-oriented supply chain. One of the first steps in the Lean journey is to identify Lean project objectives that tie to overall business strategies and objectives, and this includes metrics to measure whether or not your company is successful in attaining these objectives.

As they say, "If you can't measure something, you can't improve it." In some cases, even if you are measuring performance, you may be measuring the wrong things. Examples of where this might occur include where engineering designs products that are without a Lean supply chain in mind; accounting focuses on measures for individual processes, but does not consider the performance of the entire process; sales focuses primarily on booking orders without regard for what product mix was planned to be sold and produced; and plant management is focused on shipping dollars, efficiency, utilization, and overhead absorption metrics that go "head to head" with the goal of reducing cycle time and customer satisfaction.

Metrics Framework

In "Supply Chain Metrics that Measure Up—Building and Leveraging a Metrics Framework to Drive Supply Chain Performance," authors Faldu and Krishna point out that a metrics framework is needed that is balanced across all supply chain areas (demand planning, customer management, warehouse management, etc.). This type of framework ties the individual supply chain processes to the overall business strategy of the company. Their framework includes the following, logical steps:

1. **Establish the right metrics**—They should be reliable, valid, accessible and relevant.
2. **Link metrics to overall strategic objectives**—This is very important so that you know your supply chain is in alignment with the company's mission and strategy.
3. **Create a detailed metrics bank**—This includes a set of related metrics that relate to each supply chain process and maps the metric to the person who is accountable and responsible for its measurement and performance. [Faldu and Krishna, 2007]

Once this framework is in place, the authors point out that it is important to leverage it by using cause and effect to gain insights, such as what metrics can ensure product availability on the shelf.

Financial Impact of Metrics

Next there is a need to quantify the financial impact of supply chain metrics. For example, link cash-to-cash cycle to return on assets. This helps executive management to better understand the links between supply chain performance and overall business financial performance.

Review Scorecard during S&OP

Finally, Faldu and Krishna point out that it is important to review this scorecard as part of the S&OP process. [Faldu and Krishna, 2007]

The benefits of this type of supply chain metrics framework are better alignment with corporate strategies and objectives, better collaboration internally and externally with customers and suppliers, an increase in productivity, and greater commitment and ownership of metrics and targets.

Dashboards to Display and Control Metrics

A very common way to measure, analyze, and manage supply chain performance is with the use of a dashboard. The dashboard can be as simple as data manually collected and put into a spreadsheet with some graphs, to a more automated, visually pleasing dashboard generated by an ERP system. A supply chain dashboard helps in decision making by visually displaying in real time (or close to it) leading and lagging indicators in a supply chain process perspective.

Indicators

The metrics used in performance dashboards are typically called *key performance indicators* (KPIs). They usually fall into one of three categories:

1. **Leading indicators**—have a significant impact on future performance by measuring either current state activities (e.g., the number of items produced today) or future activities (e.g., the number of items scheduled for production this week)
2. **Lagging indicators**—measures of past performance, such as various financial measurements or, in the case of the supply chain, measurements in areas such as cost, quality, and delivery
3. **Diagnostic**—areas that may not fit under lead or lagging indicators but indicate the general health of an organization

Application Areas of a Scorecard

The dashboard, versus a scorecard, is more operational in nature and reviewed more frequently. More often than not, according to W.W. Eckerson in his book *Performance Dashboards: Measuring, Monitoring, and Managing Your Business*, dashboards are used in three major application areas:[Eckerson, 2005]

1. **Monitoring**—Make sure things are in control by watching the dashboard metrics.

2. **Analysis**—Look at performance data across different dimensions and levels to get to the root cause of issues.
3. **Management**—A mixture of performance, diagnostic, and control indicators are reviewed by executives, managers, and staff.

The dashboard allows for more granular detailed analysis, in addition to the aggregation functionality displayed in the dashboard view.

We can then conclude that it is critical to set meaningful, relevant, and attainable targets to ensure that everyone is focused on a Lean supply chain, but at the same time, be cognizant of the fact that you do not want to create "paralysis by analysis" where people end up focusing more on the numbers than on the customer.

CHAPTER 15

Education and Training: All Aboard the Lean-Train

Most of the Lean training and implementations that I have facilitated over the years have been delivered primarily by using the *train-do* method. This involves starting with some initial basic training in Lean concepts and tools. This is typically followed by team-based "critical thinking" using processes such as value stream mapping and brainstorming sessions to come up with areas for improvement. The recommended improvements are then implemented by the team using Lean concepts and tools through scheduled *kaizen* events.

This type of training, combined with my university experience as an adjunct professor, has enabled me to come to some conclusions of what works and what does not work.

There are certain training methods and tools that can be successfully applied to Lean, including its application in the supply chain and logistics function.

Training Methods

Traditional Methods

Raymond A. Noe, in his book *Employee Training and Development,* points out that traditionally there have been three major methods for training: (1) presentation, (2) hands-on, and (3) group building. The train-do method typically combines all three methods to some extent [Noe, 2002].

Presentation Method

The presentation method is where the group is passive and primarily listens to information presented to them through lectures supplemented by audiovisual means (e.g., slides and videos). This is typically one-way, from the trainer to the audience. It is a relatively inexpensive, efficient way to transfer knowledge to a large group.

This method can have variations, such as team teaching, guest speakers, and panels to make them a bit more interactive.

Hands-on Method

Hands-on methods usually require the trainee to be actively involved in the learning process. This method can be in the form of on-the-job training (OJT), simulations, case studies, business games, role playing, and behavior modeling.

OJT can be used for new or inexperienced employees who need to get up to speed on a new job, taught how to operate a piece of equipment, or for cross-training purposes. This can be taught by a current employee, by apprenticeship, or self-directed learning. These days, you do not see as many apprenticeship programs as in the past. The most common reason given seems to be that "it is hard to find good people to develop." However, qualified candidates can still be found in skilled trades and are typically sponsored by the company or union.

To be effective, the trainer should have lesson plans, checklists, procedure and training manuals, and progress report forms available at the time of training.

In this age of the Internet and Intranets, self-directed learning can be advantageous as trainees can learn at their own pace, on their own schedule, and require less supervision. However, the trainees must have a good amount of self-motivation to complete the course.

Simulation Games

Simulations are quite useful in that they "mimic" a real-life situation and the decisions made by the employee have similar outcomes to what would occur in the workplace. Simulations can be useful in developing teamwork, production, process, and management skills. In fact, in 2009, I developed a Lean supply chain and logistics management training simulation game available for this specific purpose (http://www.enna.com/lean_supplychain/).

Case Studies

Case studies are useful in showing trainees things that may happen in the workplace and allow them to develop and use "critical thinking" skills, as well as develop teamwork skills. It is not only a useful tool in the university setting, but also in the workplace as it allows the trainees, while guided by a facilitator, to see how other companies deal with business issues like theirs and then transfer some of that thinking to their own workplace.

Role Playing

Role play involves trainees acting out characters that are assigned to them in a specific scenario. The idea is to focus on responses of participants and how to deal with different situations. This can be especially useful and fun when doing simulation training games.

Behavioral Modeling

Finally (and probably least used) is behavioral modeling, in which trainees are presented with a model with specific behaviors that they attempt to repeat. This works better for learning skills and behavior more than information or facts.

Group-Building Methods

Group-building methods are used to improve team or group effectiveness. The trainees share ideas and experiences, build a group identity, and grow to understand their teammates' strengths and weaknesses.

There are a number of group-building methods. There are more adventure-type methods, such as Outward Bound, where the team goes through wilderness or outdoor training to develop teamwork and leadership skills. These methods are especially effective for improving problem-solving and conflict-management skills. The key to this type of training is a wrap-up at the end, where the results and how to apply them in the workplace are discussed.

Team Training

Team training, discussed in Chap. 10, involves getting the members to work together to reach a common goal. The optimal result of teamwork is that members learn to identify and resolve issues together. In order to be successful, the team members must be properly trained and supported by management.

Action Learning

In action learning, the team or group works together on a real issue or problem and creates an action plan to resolve it. This type of group method can be used to make changes to processes, improve use of technology, or improve customer satisfaction for example.

Selecting the Training Delivery Method

In order to determine which method is right for your company, you need to decide what type of outcome you want and, based upon that, decide which method(s) better supports that desired outcome and at what cost. Once you have that information, you can develop a training plan to support your training needs.

Consultants

While we all know the saying about consultants "borrowing your watch to tell you what time it is," we also know that they can contribute to change (sometimes radical) to a business. Most of the successful consultants these days try to get their clients heavily involved in the process in order to ensure success. If employees do not feel like they were involved and listened to, they will tend to not follow the recommendations of the consultant. Many consultants will try to get consensus by having validation of findings with executives, followed by workshops with key employees to both confirm findings and build support for implementation.

If a consultant expects to be around for the long haul, he or she will still need to do some training and facilitation of events as well. Of course, it is important to have management support (e.g., steering committee) and the assignment of Lean champions, a *kaizen* agenda, etc.

A consultant can help to solidify the opportunities, put together a rough plan, and identify the potential payoff. To really make things happen, and stick, you need your people trained and involved in the change process.

Training: Key Management Team (Seminars, Certifications, etc.)

There will need to be some level of experienced leadership with Lean to be successful. As Lean is relatively new to SCM, there are not that many people who have experienced it in this area. For the most part, a company will need to develop its leaders, or agents of change, on their own (or use employees with Lean Manufacturing experience to help in the process). The best path in this case is to select and develop key employees in your supply chain and logistics functions via external training (or possibly use external trainers brought in-house) for very specific training that can, in some cases, lead to some kind of certification or educational credit. There are many programs available by traditional methods (attending outside training programs and seminars) as well as via the Web.

As an introduction, there are many off-site (and on-site) seminars for a variety of Lean training in manufacturing, distribution, and services. These are typically good starting points to get everyone to a basic understanding of Lean concepts and can be offered to the entire company in some cases (e.g., introduction to Lean).

To truly move things forward, it can help to have key employees take courses that may involve a certification process having levels of accomplishment, such as a "green belt, black belt, etc." in Lean, Six Sigma or even Lean Six Sigma. One of the most well known is www.villanovau.com, which is a more traditional university offering various levels of Six Sigma and Lean Six Sigma certification programs. Just do a search online for "Lean training" or "Lean certification," and you will find an almost endless list of possibilities.

Once these key employees have been trained they can work with the executive team to establish training and *kaizen* event schedules for the company's SCM function.

Training: General Workforce

Just look up "workforce training grants" for your state on a Web browser, and you will find that most states have some kind of grant program for free training (typically, you will still need to pay your workers while in training). The purpose of these types of grants are to improve the skills and knowledge of the workforce and at the same time help goods and service companies to grow and prosper so that they hire more people in your state. It is really one of the best things a state can do to help businesses survive in today's competitive environment.

Most of these programs use trainers and consultants from the private sector to deliver the "free" training, but in many cases also allow for OJT delivered by qualified company employees. The grant application process and deciding which training to include can sometimes be more challenging than the actual training itself. To help with this, in many cases, there are nonprofit and for-profit companies that will assist with the grant application and management process (beware: some, but not all, of them will charge for this service).

From my experience, doing Lean training via a grant can have various outcomes. Many business owners look at this as free training and do not really have a plan to implement it. This is a big mistake and usually results in some temporary improvements but no long-term results. The management must, of course, commit to Lean transformation as mentioned earlier in the book, but also develop a plan for its successful implementation and assign responsibility for its results.

So while the trainer is both a trainer and facilitator, the company is responsible to make sure that the proper people are involved and responsible with clear goals and objectives for outcomes.

In the area of Lean SCM, especially in the retail distribution side, I have noticed that there is such an emphasis on productivity (cases per hour for example), that there is little time available for training for operators. It is critical that time is allocated in the annual budgeting process for training of this type, otherwise you will end up training small groups of people for very short bursts, which is not very effective for the long term. Again, from my experience doing Lean in the warehouse/distribution environment, it seems that very little time had been put in the annual budget for training purposes, especially training of this nature.

Obviously, you can bring in outside trainers on a self-pay basis, instead of using some kind of workforce training grant as they are not hard to find (type "Lean trainers" on any search engine). The advantage of this method is that you can have more control over the content and do not have to worry about minimum class-size, number of training hours and other restrictions that may be required by your state.

Training: Tools and Tips

Having attended a fair number of courses and seminars on presentation skills over the years has been a great help to me when teaching and training both in a university and in a business setting. One of the important things that you discover is that people learn through a variety of their senses—visual, auditory, and touch or feel. Most people learn best through a combination of all three but depending on a combination of things like education level, interest, and job requirements, the balance may vary.

Games

If you are training an executive group, you may be okay with mostly visual presentations (e.g., slide shows and some video examples). If you are working with front-line operators, who are hands-on types of people, then you will need to either have some training simulation games or actually go out into the workplace and implement change.

When you are doing an introductory-type course, it may not be possible to actually go out on the floor, so you are best suited to using a Lean simulation game. If time is limited, then something like the "paper airplane" simulation may be good enough as well as cost-effective and a great team-building exercise (e.g., flow simulation found at www.enna.com). In many cases, it may be best (especially with operations people) to use a more specific type of training simulation tool, such as the one that I specifically developed for Lean supply chain and logistics management (http://www.enna.com/lean_supplychain/).

The train-do method is very effective as it uses a combination of classroom and "on the floor" training allowing for a blended mix of visual, auditory, and touch types of learning.

Handouts and Forms

It is also helpful to never give handouts to the audience before the presentation. If you do, they tend to look down and read it while you are talking, which can distract from the learning process. You can always make handouts available to them afterwards as the handouts may include another learning tool, forms, which may be needed for the hands-on part of the training.

There are many types of forms in Lean including those used for:

▲ **Planning**—team charter and value stream map implementation plan
▲ **Gathering data**—VSM data collection form, checklists, activity and process charts
▲ **Assessing the current state of an area**—Lean opportunity assessment and 5S audits

Language Barriers

A challenge that is becoming more common these days is the barrier of language. In both manufacturing and supply chain and logistics in the United States, we find a variety of languages spoken other than English, primarily Spanish. Unless the trainer is bilingual, it is best to find someone in the audience who is bilingual and can act as a translator of sorts. It is also helpful to be able to have your handouts available in Spanish. If that is not the case, there are some Lean videos (especially on the subject of introduction to Lean and 5S) and handbooks available in Spanish (e.g., *The Lean Manufacturing Handbook* by Kenneth W. Dailey found at www .amazon.com).

In many cases, companies have a fairly large part-time workforce, which again, may be made up of many foreign language-speaking people (in the United States, Spanish is most common, of course), and detailed training may not be feasible or practical in this case. The use of the previously mentioned videos as a kind of orientation for new hires and temporary employees is helpful. The other thing you can do in this type of situation is to have plenty of standardized work, such as laminated job instructions (in English and Spanish), as well as a very visual workplace right down to marked floor assignments to make sure that work is easily understood and followed.

All of the mentioned tools and methods must be explored in order to make sure that everyone is properly trained and involved in the Lean transformation process. It may vary in each company based upon that company's specific goals and objectives.

Measuring Success

From the pure training perspective, it is always useful to survey the participants in terms of the course and trainer. I have also had clients ask that participants be tested in order to make sure that there was a clear understanding of the concepts and applications, which is not a bad idea. You can use standard quizzes or, for more fun, make a game of it. There is a Lean Jeopardy game available on www.amazon.com as well as www .theleanstore.com, which can be modified to meet your particular needs (e.g., manufacturing, office, or supply chain), as it is in slide show format.

Playing a game such as this is fun and a great way to reinforce participant learning and retention.

The benefit of the train-do method is that no matter which method or methods you use for the classroom training, you still end up applying the concepts out on the shop floor to reinforce the classroom learning and get real results with greater enthusiasm through team-building exercises like simulations and games.

Employee training (and how you go about it) is one of the keys to success in your Lean journey. It shouldn't be taken lightly and should always be part of an ongoing process, not just a onetime occurrence.

In our final chapter, we will examine the future of Lean in supply chain and logistics management in terms of people, process, and technology.

CHAPTER 16

Future Thoughts: Lean Times Ahead

The future of supply chain and logistics management as a discipline and a career is bright with ever-shortening product life cycles, continued outsourcing, growth of supply chain technology, ongoing economic challenges, and the consumer need for everything "now." However, in order to continuously improve the process by applying the Lean concepts and tools that we have discussed in this book, many things need to happen.

Lessons Learned

Some good Lean SCM lessons can be learned from the supply chain management leader's feedback in the SCMR 2007 Review discussed in Chap. 13. [www.scmr.com, 2007] Among them are:

▲ Try to make SCM an integral part of the overall business strategy.
▲ Put someone high level in charge of your supply chain, for example, a chief supply chain officer (CSCO) reporting to the CEO, to ensure top-down support.
▲ Take down any remaining bottlenecks that are hindering your supply chain advancement (both literally using VSM and figuratively by changing the culture).
▲ Intensify the focus on customer needs. Move your system from push to pull and eventually to on-demand, if possible.
▲ Use S&OP to better match supply with demand and reduce reliance on forecasts and their inherent uncertainty.

▲ Establish economic targets for supply chain improvement that match corporate goals and objectives.

▲ Create a plan for developing and including trusted business allies in building the innovative supply chain model—and share the risk.

Barriers to Supply Chain Integration

In his *APICs Magazine* article "The Chain of Alignment," John van Veen points out that "current literature focuses on five types of barriers to effective supply chain integration: technological, relationship, structure, human resource and alignment." [van Veen, 2011] Those same barriers can also apply to the successful effort for a "leaner" supply chain and logistics function.

Let's discuss some thoughts on what needs to happen on a going-forward basis within each of those categories.

Human Resources

Perhaps foremost is having capable people needed to support the function and the never ending mission to identify and eliminate waste. In the years since graduating from The Pennsylvania State University with a degree in business logistics (now "supply chain and information systems"), many other schools have added a SCM major. It is important, of course, to continue to refine the programs and make sure that students attain some level of education in Lean within the SCM context. Some universities now offer online Lean and/or Six Sigma training and certification (e.g., Villanova University offers an online Lean Six Sigma certification program). Others, such as The Penn State offer shorter executive programs, such as "Applying Lean Principles across the Supply Chain."

It is also important for companies to provide education and training "on the job" in terms of using Lean tools for continuous improvement. Some states, such as New Jersey, offer state-funded workforce training. Whether it' is supported internally or externally, it needs to happen.

In the article "Taking a Global Approach to Education," Brigit McCrea, points to a 2010 survey by The Ohio State University's Fisher College of Business of domestic and international supply chain executives found that executives are looking for "not only excellent tacticians that are good with numbers, but also are skilled at offering and executing solutions in a fast-paced, international environment. Soft skills, including those centered on communication and presentation, are also in high demand, as is the ability to function in a team atmosphere." [McCrea, 2011]

Structure

As we mentioned earlier, it is very important to have a high-level position, such as a chief supply chain officer (CSCO), give visibility to the importance of both SCM and its efficiency.

In "The Rise of the Supply Chain Officer," Carlos Gordon mentions that having a CSCO enables companies to put "constant effort into designing and managing a lean and agile supply chain that supports the company's overall strategy." This can ensure that "partners in the supply chain must be able to work both independently and together, and optimization of performance must occur at both the company and individual levels." [Gordon, 2008]

One thing we know for sure is that things change, and they seem to be changing faster and faster. As a result, it may be necessary for one company to have multiple supply chains. For example, the Heizer and Render *Operations Management* text describes how Darden Restaurants (Olive Garden, Red Lobster, etc.), has at least four identifiable supply chains (smallware, professional food distributors, independent local fresh food suppliers, and fresh seafood suppliers). Depending on the product or service this may be the most efficient way to do business. On the other hand, the individual supply chains need to be as flexible and agile as possible to react to changing customer demand patterns. [Heizer and Render, 2010]

It is also important for the supply chain and logistics function to match the company's strategy. *Operations Management* by Schroeder et al. mentions a number of strategies and how operations and supply chain must match in order to be successful. [Schroeder et al, 2011] They include:

- ▲ **Competing through quality**—Customers help to establish quality requirements, but our employees and supply chain partners make sure we meet them.

▲ **Low cost**—Focus on waste to eliminate cost that can come in terms of quality issues, as well as SCM costs in areas such as transportation and distribution.

▲ **Delivery time**—Lead time does not just come from manufacturing, but from your supply chain structure (procurement lead times, transportation modes, and distribution processing of orders, for example).

▲ **Flexibility**—Customer requirements are constantly changing in this era of "mass customization" and wanting things "now." The supply chain must be structured to accommodate this to minimize waste. If that is not your strength, then perhaps you should consider concepts such as third- (and fourth-) party logistics, postponement, etc.

In general, you may want to look at your supply chain structure to improve its operation over the long haul. The operations management text by Schroeder et al., mentions these potential improvements:

▲ **Vertical integration**—Buying suppliers and wholesalers can be an expensive proposition but can also make for a more efficient, predictable supply chain.

▲ **Process simplification**—This can involve process improvement, or even starting from scratch. No matter, the focus should be on customer requirements and shifting toward demand pull.

▲ **Change your supply chain network**—This can include changing the number and/or locations of factories, distribution centers, and suppliers. Typically, in the case of factories and distribution centers, this involves a fairly sophisticated network simulation analysis (some companies do it themselves; others use outside consultants; in either case, software is available to make the process a bit easier and more optimal). It can also involve outsourcing, such as the use of contract manufacturers to add flexibility to your network.

▲ **Product redesign**—This can involve both the number and type of products. Many companies need to do an SKU "rationalization" to reduce the number of items in their supply chain. The Pareto principle or 80/20 rule is helpful in this endeavor. Postponement further downstream in the distribution channel can be a way of simplifying the number of SKUs that you carry as well. Of course the products themselves can be modified to have more common parts (e.g., common components and modules).

⚠ **Third- (and fourth-) party logistics**—There are a variety of services that can be outsourced to 3PLs, including warehousing, transportation, and light manufacturing (usually assembly). A relatively new service is fourth-party logistics providers (4PLs) in which the 3PL activities just mentioned are outsourced and a same or different party also provides additional services such as supply chain design and planning, customs brokerage, and international trade services.

Relationships

As we discussed previously, the supply chain is actually more of a "supply web" with complex interrelationships between customers, suppliers, transportation companies, 3PLs, etc. "The bottom line is that any supply chain must be adaptable and flexible, and leanness and agility are nothing without the integration of suppliers... partners in the supply chain must be able to work both independently and together, and optimization of performance must occur at both the company and individual levels." [Gordon, 2008]

Going forward, long-term relationships with fewer, more strategic partners can create value through economies of scale and learning curve improvements. Suppliers will also be more willing to participate in Lean JIT programs and contribute design and technological expertise.

The same goes with customer relationships—at least with your major ones. The closer you can get to point of sale (POS) demand, the closer you can get to a "make what you sell" demand-pull process. So while it may be difficult, if not impossible, for companies to set up CPFR type relationships with all customers (it is best to focus on your largest customers first), more of a "partnering" type of relationship needs to be developed with customers to continue to substitute information for inventory in the supply (and demand) chain.

In this type of environment, there are challenges in controlling a supply chain involving many independent organizations. To be successful, you need to have mutual agreement on goals, trust, and compatible organizational (Lean) cultures.

Technology

As a lot of your company's supply and demand chain is external, the concept of visibility is critical in controlling waste to decrease costs, and increase efficiency and service levels.

Traditional ERP and SCM systems give you good visibility within your facility, but historically, they did not necessarily give you much if any visibility outside. Companies these days have many partners who they need to communicate with, especially with the increase in outsourcing that has gone on. So many of the big players like i2 and Manugistics (both now owned by JDA software) have worked on addressing the visibility problem with their SCM products, and others have partnered with other software companies to enhance outside visibility. The same goes for large ERP vendors such as SAP and Oracle.

Supply chain visibility can be looked at from either a logistics perspective (technology that tracks a product and materials as they move through the supply chain) or a solutions perspective that enables manufacturers to track products throughout the various stages of its lifecycle.

A supply chain benchmark survey by Gatepoint Research and www.E2open.com said that greater than 50 percent of the respondents had in excess of 500 component suppliers or manufacturing partners with 44 percent having poor visibility into tier-1 suppliers, and 75 percent had poor visibility into tier-2 and tier 3-suppliers. In all, greater than 80 percent stated that they had not automated, or only partially automated, their supply chain processes. [www.E2open, 2009]

A 2010 Aberdeen Group survey found that 57 percent of the respondents felt that improving supply chain intelligence was a critical factor for improving overall operational performance. [www.aberdeen.com, 2010]

According to Mary Shacklett in her article, "Supply Chain Software— The Big Spend [Shacklett, 2010] there are three major software trends:

1. There is a move away from traditional corporate thinking that a high degree of software customization to the business produces competitive advantage. Instead, contemporary thinking is that packaged software in a "vanilla" form can do an adequate job for the supply chain, as long it incorporates industry-wide best practices. Customization for corporate business processes still occurs, but it occurs via technologies like services oriented architecture (SOA), which splits off pieces of business logic that can be assembled to support any end-to-end business process without altering the core software. In this way, companies remain eligible for new software releases (and support) from their vendors and also benefit from the R&D the vendor puts into the product to continually improve it.

2. More built-in capability for analytics that assist managers at different levels of an organization with visibility, reporting, metrics, and analysis of what is (or isn't) going on in the supply chain. Software development is focusing on increased intelligence embedded in the software, as there is a demand for visibility, which can facilitate the accuracy of forecasts.
3. Migration to cloud-based supply chain solutions. This is largely being driven by a need to get a handle on external company transactions with partners and suppliers. [Shacklett, 2010]

Alignment

The concept of alignment refers to having all of the members of your supply chain all moving the same direction so that the entire supply chain can be Lean and flexible.

In the "The Chain of Alignment," John van Veen makes the case that "inconsistent goals challenge successful internal and external supply chain integration. Divergent objectives lead managers to make self-interested suboptimal decisions that frequently are in opposition to those of other business managers and supply chain members." [van Veen, 2011] The net result of this type of activity is waste.

A 2009 study for the CSCMP entitled "Mastering Supply Chain Management" by Fawcett et al. found that the largest barriers to the collaborative business model were a turf protecting organizational structure, resistance to change, conflicting measures, lack of trust and lack of management support. In other words, pretty much the same barriers to lack of success implementing Lean in the supply chain. [Fawcett et al., 2009]

Some of the potential solutions to breaking down these barriers included:

▲ Collaborative strategy meetings
▲ Executive steering committees
▲ Collaboration workshops
▲ Cross functional teams
▲ S&OP
▲ Co-located managers
▲ "C" suite SCM executives
▲ Supply chain advisory boards

It goes without saying that a Lean supply chain journey requires an executive steering committee and cross-functional teams, but the most intriguing and potentially most effective concepts mentioned were:

▲ Supply chain advisory boards made up of representatives of key customers and suppliers (and perhaps current software vendors, 3PL providers, public warehousing, and transportation partners)
▲ Co-located managers, which is a concept that larger companies can afford to do (e.g., Proctor & Gamble had up to 200 employees on site at Walmart Headquarters for a VMI program at one point in the 1990's)

Both of these concepts can make major improvements in visibility and collaboration.

Trends in Lean Supply Chain

As stated at the beginning of this book, Lean has gradually migrated from various types of manufacturing processes (it started in repetitive assembly-line processes, such as the auto industry, and moved to continuous and batch processes, such as the chemical industry) to the office and now, more recently, to the supply chain.

It is even being implemented today in service industries, such as hospitals, hospitality, retail, and restaurants. If we look to the near future that trend will not only continue, but likely accelerate with shortening product life cycles and lead times, continued global outsourcing, and increasing demand for mass customization.

Complementary to this, we can see a definite trend toward more visibility through the entire supply chain, collaboration and connectivity, warehouse automation, electronic data capture, and global tracking of goods movement, which will contribute to the reduction of waste in the supply chain.

All of these trends generate massive amounts of data that need to be captured and analyzed.

Data Analytics

There is a growing trend in the use of *data analytics* (DA), which is the science of examining raw data with the purpose of drawing conclusions about the information. This information can help to gain better visibility and improve collaboration between supply chain partners, ultimately enhancing or creating value and reducing waste.

The article "The Data Analytics Boom" by John Jordan in a 2010 issue of *Forbes Magazine* points out that there are many reasons for the rise in the interest level in DA. They include:

- ▲ A generation of managers trained in the tools of Six Sigma and TQM has learned the value of the application of data to improving a process.
- ▲ An explosion of available data and capabilities to process that data.
- ▲ Many of ERP systems have recently put added emphasis and capabilities into analytics.

Supply Chain Analytics and Lean

We discussed the use of Lean analytical tools in Chap. 6, which can be applied to the supply chain.

Data analytics is especially important in the supply chain and logistics field. At The Pennsylvania State University, which has one of the leading supply chain programs in the country, one of the primary learning goals in their masters of professional studies in supply chain management program is to "apply supply chain analytics to improve operational effectiveness and/or make the supply chain a source of competitive advantage" (source: www.smeal.psu.edu).

In "Fueling Supply Chain Transformation—Predictive analytics energizes dynamic networks," Dogan et al. point out that the increased complexity, shortening product life cycles, amplified price, and volume volatility require managers to have real-time insights into the impacts issues like these will have on the supply and demand sides. [Dogan et al., 2011]

Advances in technology have allowed for real-time analytics to give management more visibility and deeper analysis to be ready for these volatile and ever-changing conditions with more realistic strategies.

This is not all theoretical either. The article points out a real-life example of a major agribusiness company that wanted to improve its supply chain performance. They used supply chain planning analytics to identify the root causes of performance problems and apply the findings to their forecasting, inventory, and delivery functions. The results have included a

20 percent reduction in inventory and working capital, a13 percent reduction in transportation and distribution costs, and an 8 percent reduction in the cost of goods sold.

The main point here is that Lean supply chain transformation programs can benefit from predictive analytics, and this task is becoming easier with the ever-increasing amount of available data and user-friendly tools to gather and analyze it.

Potential Obstacles to Lean Thinking in the Supply Chain

There are a variety of opinions as to the obstacles to Lean thinking in the supply chain. Tom Craig, president of LTD Management, feels that the most significant obstacles are "accounting, the "four walls" mentality and not focusing on the international side of the inbound supply chain." [Craig, 2011] Kelly Thomas, senior vice president of manufacturing at JDA Software Group, Inc., feels that "the two greatest obstacles to lean thinking are executive leadership and understanding that the journey never ends—in other words, sticking to it once the journey starts. Technology is instrumental in helping with the second of these. Technology can institutionalize lean techniques and provide staying power. Many lean implementations have depended on individuals and manual techniques and if those individuals change positions within the company, much of the enthusiasm and leadership are lost. Some technologies can provide interlocks similar to the concept of poka-yoke, ensuring lean techniques have staying power." [Thomas, 2011]

No matter what the obstacles, if the challenges are met with a sound methodology and commitment as has been laid out in this book, then failure will not be an option.

Lean Ahead

This book was intended to help supply chain and logistics professionals, corporate executives, teachers, trainers, and consultants to use Lean tools and applications to navigate through the complex and ever changing supply "web."

When looking at the challenges ahead, you should always be looking to identify and minimize waste wherever it may appear in your supply chain and logistics function. It is always important to realize that, as in Lean manufacturing, not all companies and industries are the same. As a result, Lean tools used will vary. It is always a mistake to use a "cookie cutter" approach, which may doom you to failure. It is also important to have top management support, plenty of training (and incentives) for everyone, and a truly Lean culture to make sure that you have a Lean *journey* and not just some training or a project or two.

The road is littered with companies that did not change with the times, but their competition did. You should think of Lean as a competitive tool that will protect you from this fate and enable you to get and stay at the front of the pack.

The slides available at mhprofessional.com/myerson and the Lean Assessment Scorecard in Appendix B should help you in this quest. Good luck in your journey and stay the course.

Real-World Examples of Lean Supply Chain and Logistics Management

The following Lean case studies are meant to give readers real-world examples of companies that have used various continuous improvement tools to enhance their supply chain and logistics functions. In keeping with the general theme of the book, they are organized using the SCOR model of Plan, Source, Make, Deliver, and Return.

PLAN

Lean Case Study: The Organized Office*

Three people worked on the same task at a cluster of desks sharing a printer. Every computer transaction generated a print and after about 10 transactions, one of them would stand up, go to the printer, sort out their own work from the printer output tray and return to their desk.

When the staff reviewed the layout, they recognized that time was spent walking to and from the printer and sorting through all the print-outs to find their own, so they came up with this revised layout.

Simply by turning the printer 90 degrees means that one person can now sort through the printer output without moving from their chair. As that person sorts through the print-outs, they hand on the work belonging to their two colleagues. And the end result? No more walking to the printer at all, and only one person rather than three people sorting through the work.

This simple measure—moving a printer—created an increase in throughput of more than 5 percent.

---

*Source: www.obviousoffice.com

PLAN

Lean Case Study: A Lean-Six Sigma Duo for the Office*

The experience of a European life insurance provider highlights the lessons learned from transferring Lean from the shop floor to the office, as well as providing a deployment model that integrates Lean, Six Sigma, and process management.

Combining Lean tools and the Six Sigma methodology has become popular during the last 5 years. However, most of these efforts were focused on manufacturing operations. The experience of a European life insurance provider highlights the lessons learned from transferring Lean from the shop floor to the office, as well as providing a deployment model that integrates Lean, Six Sigma, and process management.

A First Attempt at Integrating Lean and Six Sigma

A leading life insurance company decided to implement a comprehensive process excellence program, using Lean and Six Sigma to achieve substantial cost savings. With aggressive cost reduction goals, the Lean aspect of the program was considered the more critical to demonstrate credibility. As the first Six Sigma project can take between 6 and 9 months to complete, Lean efforts (through 1-week bursts of activity called *kaizen*) often take only weeks to deliver substantial improvements.

Having conducted some initial analysis, the team in charge of the overall effort decided to conduct a Lean pilot project to test the assumption of early wins. Using an experienced instructor who had used Lean extensively in manufacturing settings, a group of candidates was pulled together for 1 week of Lean training using traditional training modules.

While the training was well received, the *Kaizen* workshop revealed a number of issues:

▲ Many of the tools could not be applied.
▲ Changes could not be implemented within 5 days.

---

*Source: www.isixsigma.com; Article courtesy of Rath & Strong management consultants.

▲ Ideas generated by the team had to be vetted by the managers in charge of the process.

▲ Estimates of the potential cost savings were too high, causing management to lose interest.

▲ To achieve the targeted savings, the scope would have to be increased from a single process to the entire value stream.

▲ The Lean pilot area was not strategic and not aligned with the Six Sigma project selection effort.

Refining the Approach

The lessons of the early pilot program helped to refine the approach to deploying Lean in the office. The first intuitive step was to eliminate tools that could not be applied from the training program. Even more significant, a process management framework was added to the program as the organizing framework to identify potential projects and to sustain the improvements achieved. Finally, the Lean curriculum was integrated into the DMAIC framework to emphasize the message that the company was using a common problem-solving approach, with Lean tools focused on reducing cycle time and Six Sigma tools aimed at reducing defects.

The new approach to deployment was focused on process ownership: An initial two-day workshop with the leadership team responsible for a business process focused on mapping the entire value stream and identifying a set of key metrics. This workshop was followed by a period where a team of Black Belts validated the process map and collected data. Then, a second management team workshop reviewed the initial dashboard, developed a future state map and identified a list of high-impact projects that would be necessary to accomplish the future state. This approach ensured that all process improvement projects had a strategic focus and were based on a real business needs. Finally, the leadership team reviewed the list of projects and identified which methodology (Six Sigma or Lean) should be used to address each.

Once the Lean projects had been identified, internal facilitators went through a 5-day training program to become Lean experts. In this role, they helped select the team for the *kaizen* event and worked with the leadership team to ensure that the team charter was narrowly focused on one specific measurable goal.

The *kaizen* event itself was split into three parts:

▲ **First *Kaizen* Workshop:** A 2-day workshop was used to define the problem, collect data, analyze the issues and generate ideas on how to tackle these issues.

▲ **Validation:** The initial workshop was followed by a 2-week period during which the Lean specialist with the team validated the assumptions behind its ideas, collected additional data and ensured buy-in from the management team.

▲ **Second *Kaizen* Workshop:** During a 3-day period, the team reviewed the additional data, developed an implementation plan and completed the new process design.

Moving away from the typical *kaizen* approach was critical to ensure that management and colleagues were committed to the project, as well as to avoid overestimating the impact or underestimating the effort required to implement.

Lessons Learned

The initial projects generated valuable insights into what is different about applying Lean in an office environment.

1. **Quality of the Inputs Is Critical:** While quality of the inputs is obviously a concern in manufacturing processes, in service businesses the importance of getting customers to provide correct and complete information is crucial. In a manufacturing environment, one can typically assume near defect-free inputs. For a service business where the customer is the supplier, substantial error rates are typical. In one case, only a third of all incoming applications were accurate and complete. This results not only in substantial rework but also causes enormous delays and increases the number of items in the workflow, which in turn results in backlogs, expediting and numerous other wastes. Linking a Lean effort to streamline the process flow with a Six Sigma effort to improve input quality can be very powerful. Even without trying to improve the quality of the inputs, simply by separating those items that are ready for processing from those that require investigations and rework early on, significant results can be obtained: Incomplete applications no longer clutter the downstream processes.

2. **The First Process Step Is the Pacemaker:** One of the fundamental ideas of Lean is the concept of takt. Takt time is the rate of customer demand, and in a manufacturing environment this concept is typically used to smooth production. Manufacturers have significant control over their inputs and can instruct suppliers as to when to ship how many parts. In services, this concept is more difficult to apply, since in most cases the customer also is the supplier and the company has little or no control over the volume and sequence of inputs. With uncertain demand and a steady capacity, this often results in significant wait times for customers. Applying the Lean principles to a service process requires either stabilizing the flow of inputs or creating flexible capacity. Either way, where a Lean implementation typically starts with the process closest to the customer, for service businesses it typically means starting with the processes closest to the supplier.

   Streamlining the work flow of all subsequent processes and stabilizing the process requires the first process to act as the pacemaker and to release new work in fixed quantities. For example, by releasing 60 applications an hour every subsequent operation now has a predictable volume to work with at any given point in time.

3. **Cross-Train and Establish Teams:** Few service businesses have a rigorous training program for the rank and file. In industries such as banking and insurance, few employees are able to handle a variety of tasks. Learning happens mostly on the job, and the turnover rate makes training a poor investment. Typically, jobs in a service business are defined rather narrowly to allow rapid training and easy replacement. As a result, flexibility suffers: employees cannot easily shift from one job to the next. That creates complex staffing problems and makes it extremely difficult to adjust to changes in the mix of business.

4. **Design for Flexibility Is Key:** Service businesses do not have the luxury of finished goods inventories as a buffer for the inherent variability of demand. And although they can stockpile work in process (backlog) and use it as a buffer to keep everybody busy, this inventory is extremely expensive:

   ▲ Customer satisfaction suffers due to long response times.
   ▲ Customer contact centers receive multiple inquiries from customers who want to know when their application will be processed.

▲ Systems are needlessly complex as they must be designed to hold and store backlogs.

▲ Management systems are needed to track the status of individual tasks.

Overall, local efficiency soars while global effectiveness suffers. An effective Lean deployment needs to consider the entire process from beginning to end.

5. **Dependency on IT:** Lean earned its reputation for speedy results on the shop floor, where it was possible to have Lean teams rearrange machines to achieve flow and limit inventory to a minimum. *Kaizen* events are designed to generate enthusiasm and momentum by encouraging teams to make changes as they go through the event, keeping a daily scorecard that tracks progress toward goals such as cycle times, inventories, etc. Using the same techniques in a modern service business, where workflows are driven by IT systems and the majority of work happens in the system can lead to disappointments: improvements take longer to implement, because they require changes in the wiring of the workflow. In these situations, a standard Lean approach will simply not yield the desired results.

Teams must have access to IT resources to ensure that necessary system changes do not end up in the long list of IT backlog but can be addressed outside the normal process.

Conclusion: Integrate into Process Management Framework

While application of the Lean tools and concepts can have a significant positive impact on service businesses, the tools need to be adjusted to reflect the differences that exist between manufacturing and services. Even more important, the effort should be integrated into a process management framework to ensure that the Lean projects tackle the key issues.

PLAN

Lean Case Study: Heinz Gets to One Number Forecasting*

Food Giant Makes Major Supply Chain and Financial Improvements as It Professionalizes the Forecasting Discipline

While many companies have used Sales & Operations Planning and related strategies to develop a consensus or "one number" forecast, many others are still trying to reach this goal.

Food giant Heinz is one company that has moved well down the path of forecasting excellence, according to Sara Park, Sr. Manager of Forecasting and Demand Planning at Heinz.

Writing in the most recent issue of *The Journal of Business Forecasting*, Park says that in the early 2000s, forecasting was formally the responsibility of brand management but, in practice, many functional groups (finance, manufacturing) each maintained their own separate forecasts.

As a result, Park says, "Everyone, especially Brand Management and Sales, ended up spending too much time debating "true demand." Different groups were developing different forecasts based on use of different assumptions."

However, the company was moving to professionalize its forecasting process and team. A group was developed, first under marketing, to improve the process. A key first principle: fact-based management. "We insisted that forecasting volumes (were) based on solid facts and data with a 100 percent dedication to accuracy," Park said.

Until 2004, that small forecasting team was using basic tools: spreadsheets, AC Nielsen data, and shipment data from internal systems. Heinz then implemented a new demand planning system (from JDA Software), at which time the forecast team moved under the VP of Supply Chain.

---

*Source: Supply Chain Digest, October 20, 2008.

From Top Down to Bottom Up

Perhaps the key process change in the Heinz journey was the switch to a "bottom up" forecasting approach. Previously, the forecasts started with a top down approach based on brand targets. It also relied heavily on static allocation of forecast volumes within business segments. Item level forecasting was "something of an afterthought." For example, in the pasta sauce category, the critical SKU level forecasts were divided among different flavors based on historical percentages—which overlooked multiple factors and different growth trajectories among the different SKUs.

Park says Heinz made two critical changes:

- Forecasting consumption first, then shipments; and,
- Forecasting at more granular level, so that each SKU was handled properly for growth, promotions, etc.

Heinz also began to do a better job of forecasting by channel. This was critical because some channels, such as warehouse clubs, were growing faster than traditional channels, and had opportunities for additional growth. In the end, Heinz wound up with a brand-channel-fiscal month/quarter/year rolling forecast, providing significant insight to marketing, sales, finance and the supply chain.

The single forecast put the entire company on the same page. Planning and budgeting became more efficient and less contentious, as the same "volume call" drove every department's own plans. When spending increased or decreased (for example, in the promotions area), "appropriate volume was either added or subtracted from the forecast." When large customer events were added or moved, volume was correspondingly shifted.

The effort also had big supply chain benefits. Planners and managers were able to see potential supply issues or volume spikes much earlier, and work with vendors proactively to resolve them.

Accuracy Takes Center Stage

An important change came when Heinz's CEO stated that forecast accuracy was now of paramount importance. In the past, "under promising and over delivering" was often considered a commendable approach. This type of CEO endorsement often also makes it clear that "hiding" bad news is not acceptable.

The one number change is not without challenges, of course. It puts a lot of pressure on the demand planning function to acquire and process data from many sources, and to hold the whole process together for the entire company. Forecasts are, as always, subject to error, and often major, long-term strategic decisions and investments are made on those forecast numbers. Ironically, demand planners have to sometimes really struggle to meet the needs of the supply and demand sides of the house—issues never really created in a world of multiple forecasts. Still, Park notes, "One-number forecasting has to be a principle that we want to live by."

SOURCE

Lean Case Study: Automotive Lean (Outsourced) Supply Chain*

Background

An automotive company wanted to investigate the opportunities for cost savings by outsourcing the production of a module of a vehicle. Outsourcing manufacturing work can significantly reduce direct labor cost. However, if the supply chain is improperly designed, it could increase costs in shipping, handling and inventory, which can outweigh any potential savings.

Based on the client's own preliminary analysis, they believed that outsourcing the production would probably be cost advantageous. In the detailed analysis, the client would like to examine several proposed options for outsourcing to determine which option would be the best to minimize inventory and transportation costs and while maximizing responsiveness and flexibility.

Methodology

The key investigation items are:

- Which supplier(s) should produce the modules?
- What is the optimal production/shipping process?
- What operating patterns should be followed by each supplier?
- How much inventory does each supplier need to hold?
- How many dedicated shipping docks at the assembly plant and at each supplier will be required?
- How many trucks will be needed to ship the finished modules?

The steps in the study are:

- Calculation of cost savings from material handling, floor space requirement, tooling and storage

---

*Source: www.georgia-associates.com

- Calculation of manpower cost savings from direct labor such as material handling operators
- Analysis of the impact of operating pattern, transportation, and inventory on costs

Results

After the best supplier and optimal logistics configurations have been identified, the key findings were:

- By outsourcing, the client would save U.S. $2.15 million annually in direct employment costs.
- Another $1.5 million can be saved in areas such as fuel and vehicle maintenance.
- However, an extra $1.4 million (annualized) would have to be incurred to install new equipment on the supplier's premises.
- Changes to other cost components would be negligible.
- The client can therefore expect approximately $2.25 million in net annual savings by outsourcing the production of the module.

SOURCE

Lean Case Study: Cessna E-Procurement Project*

Cessna E-Procurement Initiative

Cessna was a traditional, conservative culture when management announced a supply chain initiative they hoped would revolutionize purchasing processes and give them a competitive edge. Dave Oppenheim, then director of e-business, realized that electronic data interchange systems were critical for the company-to-company integration Cessna hoped to accomplish.

Cessna wanted to reduce or eliminate many purchasing-related expenses and to free up overworked buyers for more strategic tasks. To avoid upfront costs and development delays, they decided to outsource.

Recognizing that supplier acceptance is key to the success of any e-procurement system, Cessna looked for a system that was supplier-friendly and would not reduce their costs at the expense of pushing them back on suppliers. They also realized they needed a system that could accommodate all sizes of suppliers regardless of their technological capabilities.

In 1998, Cessna selected ESIS because of its large supplier base and its reputation as the largest working aerospace exchange. ESIS worked with Cessna and its 1,200 suppliers to get the system up and running in less than four months.

Since the beginning, ESIS and Cessna have worked together to address Cessna's expanding procurement needs. Since the original implementation, inventory advice and shipment schedule modules have been added to fulfill Cessna's receiving requirements. When Cessna implemented Ariba Buyer in 2000, they decided to use ESIS to deliver the purchase orders to the suppliers, enabling them to conduct business the same way for both direct and indirect procurement. As part of the Ariba implementation, ESIS added the capability to create a "spontaneous supplier" for Cessna, allowing them to send "spot buy" orders for new suppliers through ESIS.

Source: ESIS, Inc. www.esisinc.com

Today the Cessna division of Textron is considered a model for e-procurement and has been featured widely in technology and supply chain management journals and at national conferences.

Value Proposition

- ▲ Provide Cessna with supply chain management technology to advance its lean manufacturing objectives.
- ▲ Greatly reduce time spent on purchasing while providing additional ROI from eliminated manual processes.
- ▲ Overhaul the buyer/supplier communication process, making it interactive and more efficient.
- ▲ Provide necessary information to suppliers automatically on an as-needed basis.
- ▲ Improve JIT inventory processes.

Challenges

- ▲ Reduce and eliminate costs.
- ▲ Eliminate manual systems.
- ▲ Free up personnel for strategic tasks.
- ▲ Integrate suppliers into business processes.
- ▲ Accommodate all sizes of suppliers.
- ▲ Implement rapidly.

Solutions

- ▲ Implement HOM system.
- ▲ Communicate with suppliers online and via EDI.
- ▲ Automate purchasing processes.
- ▲ Share planning schedules with suppliers.
- ▲ Automate advance ship notification and invoices.
- ▲ Notify suppliers of approved ship dates.
- ▲ Deliver POs for indirect purchases from Ariba Buyer.

Benefits

- ▲ Fax, phone, and paper orders eliminated.
- ▲ Supplier on-time delivery improved.
- ▲ All suppliers accommodated.
- ▲ Re-keying of data eliminated.
- ▲ Orders/Change Orders, Acknowledgements automated.
- ▲ Time to process PO greatly reduced.
- ▲ Purchasing personnel freed up for more strategic assignments.
- ▲ Purchasing personnel numbers remained flat while company production increased by 250 percent.

MAKE
Lean Case Study: Lean Kitting*
Overview

Kitting is the first step in printed circuit board assembly. It is initiated well in advance of the actual production start to be able to prepare and deliver the kit on time. Kitting involves the gathering of all the parts needed for a particular assembly from the stockroom and issuing the kit to the manufacturing line at the right time and in the right quantity. This paper discusses kitting, describes ways to eliminate waste in different phases of kitting, and illustrates lean kitting using a case study conducted in a major contract manufacturer site.

Introduction

Traditionally, kitting is initiated by the production control department based on the shop floor order as generated by the plant's ERP/MRP system. Production control will first verify that adequate quantity is available for each part number. If there are parts shortages, parts are ordered. In general, the kit is not released to the stockroom until parts arrive, but in some special cases a shop order with a part shortage can be processed. Production control then releases the kit to the stockroom for picking. The kit is typically sent to the off line setup area within 48 hours. The time it takes to pick all parts depends on kit size and number and skill of employees.

When executed properly, benefits of kitting include:

- Maximize value add time of operators
- Easier operator training resulting in reduced training cost
- Maximized machine utilization—no line stoppage due to part shortages or searching for parts
- Reduced WIP
- Reduced lead times
- Reduced part damage due to excess handling

*Source: www.optimalelectronics.com; written by Ranko Vujosevic, Ph.D. Optimal Electronics Corporation, Jose A. Ramirez, Larry Hausman-Cohen and Srinivasan Venkataraman, The University of Texas at Austin Department of Mechanical Engineering.

ERP/MRP systems are inherently inaccurate at maintaining on-hand inventory quantities as material is moved to and from the shop floor during kitting and restocking processes. Worse, ERP systems typically only know the total quantity of a part type and not how that material is delivered (i.e., 20,000 total components vs. 4 reels of 5,000 components). This inaccuracy and lack of granularity, combined with stockroom personnel mistakes, lead to kitting problems that include:

▲ Insufficient quantity of components
▲ Excessive quantity of components
▲ Wrong components
▲ Incomplete kits
▲ Insufficient quantity of component packages (for example, insufficient quantity of reels for split parts)

These kitting problems will lead to increased machine downtime, lead times, and manpower.

Various shop floor supply schemes that vary the location of component storage units with or without kitting are used by electronic assembly plants. The following cases are the most typical:

▲ **ERP driven kitting from central stockroom.** Production Control releases kits generated by the ERP system to the central stockroom. Production Control relies on the ERP system to ensure enough quantity is available in the stockroom. Inaccuracy of on-hand inventory counts in the ERP system is a major issue in this approach. In case part shortages occur during shop floor order runs, the operator must walk to the stockroom, fill out paperwork for additional parts, wait for parts, and then walk back to the assembly line to do changeover and continue production. This can take 15 to 45 minutes and represents waste and lost production. This approach is still very much in use, especially in large contract manufacturing sites.

▲ **ERP driven point of use kitting.** This process is similar to central stockroom kitting, except inventories are located in storage units located on the shop floor close to the assembly lines. Storage units can be dedicated to a customer, to all SMT lines, to a set of assembly lines (for example, lines that place components on one side only), or to the entire shop floor. The advantage of this approach is reduced machine downtime and reduced stockroom manpower. The disadvantage is increased floor space requirement and an increased level of control required to ensure operator discipline in picking components. The issue of inaccurate on-hand inventory counts still remains. This approach is more likely to be found in OEM assembly plants.

⚠ **In-house controlled supermarket-based kitting.** Components are received and stored into the central stockroom. Supermarkets are placed throughout the shop-floor. The daily demand of parts and supplier lead times are used to determine the optimal inventory levels. *Kanban* cards are used to replenish supermarkets. Dedicated personnel are responsible for collecting and converting *kanban* cards into replenishment orders which are sent to the central stockroom. Kitting is performed directly from supermarkets by line operators. The advantage of this approach is reduced manpower. The disadvantages are operator training, lost inventories (if a reel is misplaced, a new one is ordered without locating lost one), and complex replenishment logic that requires advanced inventory control software. The issue of inaccurate on-hand inventory counts still remains because counts are still maintained by the ERP system. A variation of this approach is where only the most frequently used parts are stored in supermarkets and kitting is a combination of central stockroom kitting and supermarket kitting.

⚠ **Supplier controlled supermarket based kitting.** A central stockroom is not used. The ERP system-based component procurement is abolished. A contractor who can supply all types of components is used. On-hand quantity requirement for each component is established based on the ERP forecasts (weekly consumption), or even better, actual sales. Part consumption and attrition is maintained using an MES system. The component supplier gets (daily) data from the MES system on part consumption. The supplier restocks parts once a week based on MES supplied data. The advantages of this approach are reduced inventory cost, more accurate inventory counts, reduced manpower, reduced paperwork and purchase order cost. The disadvantage of this approach is initial implementation cost. Also, tying all inventory needs to one or a few suppliers may be risky.

⚠ **Outsourced kitting.** Component storage and kitting are pushed to the suppliers, as done by Toyota in the San Antonio Tundra plant. This is a way of reducing manpower and pushing waste to the suppliers. In the Toyota plant, suppliers deliver parts just-in-time and perform kitting for assembly lines. Toyota operators just assembly parts, which maximizes operator value added time and reduces manpower for Toyota. The same approach of outsourcing component storage and kitting is available to electronics assembly companies. Outsourcing kitting will result in a reduction of purchase orders, reduced administration costs, cash flow benefits from reduced inventory, reduced manpower, and reduced plant space requirement. Again, the disadvantage is the risk of loss of control of the materials management process.

Is Kitting a Waste and Should It Be Eliminated?

In a lean world, kitting is considered a waste. Following Toyota's lean thoughts of the past, we should provide component storage racks, replenished using *kanban* signals, alongside assembly lines and have operators pick parts to set up the job when needed. In addition to that value added work, operators will also be responsible for locating parts, performing part verification, setting parts on feeders and setting feeders on machine. This approach, although eliminating kitting as a waste, may create waste in the area that does much more damage to the company's bottom line—assembly line throughput and process quality. An attempt to eliminate kitting should not be undertaken without much consideration of how it may affect the overall process and assembly line efficiency.

Furthermore, it has been recently reported that Toyota has started using kitting in some of their plants for high volume assembly operations [Lean Enterprise Institute, 1997]. Toyota has implemented a new kitting process, called Set Pallet System (SPS), in their new production facility in San Antonio that makes Tundra full-size pickup trucks. Previously, lineside storage racks were used by operators to pick parts. Operators would walk from their assembly station to each rack and pick parts to install. SPS introduces kitting personnel that receive a signal with a list of parts to be kitted, pick parts from storage racks, and then deliver pallets of parts to the assembly stations. Assembly operators are not involved in the part picking process any longer. The advantage of this approach is more value added time by the operators, cleaner work areas with visual control, fewer part selection errors, and easier training of assembly operators. The disadvantage is increased manpower by adding kitting personnel.

This approach is equivalent to the supplier-driven supermarket kitting for PCB assembly, as explained in the previous section.

Lean Kitting

The approach taken in the kitting process improvement project described in this paper can be summarized as follows:

- ▲ Eliminate waste related to machine downtime caused by invalid kitting
- ▲ Kitting done right the first time
- ▲ Eliminate waste in kitting

The priority is to eliminate waste on the assembly line by making sure machine downtime due to kitting problems does not happen. Next is to eliminate waste and make the kitting process as lean as possible.

This Lean Kitting project was implemented at a large electronics contract manufacturer's site. The project involved an assembly line that included a new Fuji NXT pick-and-place line. The NXT machine is based on a new concept of modular, scalable, and reconfigurable pick and place machines. This NXT machine had 10 modules. Thus, the same part number may occur on different modules as a result of placement sequence optimization and load balancing.

In consultation with the plant management, the following goals were set:

- ▲ Reduce kitting cycle time.
- ▲ Reduce manpower.
- ▲ Reduce number of partial material packages returned to stockroom.
- ▲ Eliminate the issue of insufficient quantity of material packages (reels) for parts split between different modules on NXT machine.
- ▲ Eliminate the possibility of wrong components being kitted by implementing electrical component test and component verification.

Kit quantities per part number calculated by the ERP system do not take into account that a part could be split between a number of modules during machine optimization. For example, if the kit requirement for a part that is split between two modules on a machine is 4,500, the stockroom may find a reel with 5,000 parts and kit it for that part, which will cover the BOM quantity plus estimated attrition. The single reel will be sent to off-line setup and it will need to be split, or another reel ordered, which delays off line and adds manpower related costs. Another case is when the stockroom does kit two reels, with quantities of 4,000 and 1,000. Let us assume that the quantity placed from the first module is 2,800 and from second module is 1,700. The off line setup personnel will have two reels to prepare for the run and production can be started, but there is not enough quantity on the second reel and the machine will stop until a new reel is provided, adding to the machine downtime.

In this project, all activities that lead to delivering the kit to the assembly line, were considered to be part of kitting, including:

▲ Pulling enough quantity to place each part (including attrition)
▲ Determining which reel will be used first
▲ Verification of component electrical characteristics
▲ Verification of component feeder type, feeder rotation, and height verification
▲ Lead-free compliancy
▲ Setup verification

Even though kitting can also include delivering the tooling kit to the line, it was considered out of the scope of this project.

Pre-improvement State

The line kit was prepared by stockroom personnel based on the kit released by production control. In case the kit was short, the stockroom would hold it until missing parts were provided. The line kit contained parts for all machines in the line. The parts needed to be separated by machine and by first and backup reels. For the NXT lines, the kit was separated by module by the operator visually checking license plates (also sometimes referred to as ReelIDs; a unique barcode identifier applied to each piece of material) and comparing them and their part numbers with the setup sheets. Separation of the NXT kit across modules took on average of 2.5 hours for the bottom side and 1.5 hours for the top side. For the NXT line with 10 modules, the kit was separated into 3 bins with 3 separations per module each, and the license plates for the 10th module were put into the last bin and marked as the 10th module.

The NXT kit was then moved across the plant to the NXT off line setup area. First, an operator was dispatched to start the electrical component verification, using an LCR meter (Inductance, Capacitance, Resistance). A second operator was sent after the first operator was given some time to test a certain amount of license plates. The second operator scanned license plates that were already LCR-verified and married them to feeders (via bar-coded feeder IDs) to complete offline setup verification. Component verification using an LCR meter was performed for the first license plates only and not for the backup license plates. Verification took on average of 55 seconds per component. Each license plate was tested, even if it had been previously tested. After the test, a "pass" sticker was put on the license plate. For components that could not be measured, license plates were put aside and another operator would verify those license plates visually, using component markings or manufacturing part numbers.

After the first license plate was verified on the feeder, it was placed on the feeder cart. After the entire setup was verified, feeder carts—one for each NXT module—were moved to the "NXT Supermarket/Kanban" area to wait for the online machine setup.

It took 3 people an average of 6.8 hours to process the kit that had been received from the central stockroom and deliver it to the NXT line. For components that had more than one component package kitted, it was important to maintain the order the packages were mounted on the machine, to minimize the number of partial packages that needed to be returned to the stockroom after the job was completed.

In the remainder of this paper, the term "license plate" will indicate a label attached to each piece of material that has a material ID and part number bar-coded on it.

After analyzing the process, the following types of waste were identified:

Extra Processing Steps

- Manually separating materials among NXT modules.
- Re-ordering of part numbers that go into multiple NXT modules.
- Redundant reel component testing which was separate from setup verification.
- A "pass" sticker placed on the license plate when component has passed electrical test.
- Splitting license plates if that is chosen as the solution for the ERP system not supplying enough quantity of license plates for split part numbers.
- Difficulty locating components that are separated from license plates for electrical test since the company used black static-free mats for placing components.

Defects

▲ License plates that fail electrical test were placed into "pass" bin and later used in assembly.

Motion

▲ When scanning feeder, scanner label was on the opposite side of the feeder compared to the operator, forcing the operator to reach over and twist arm and hand to scan bar-coded feeder ID.

Waiting

▲ If ERP system did not supply enough quantity of license plates for split part numbers they were ordered from the central stockroom.

Transportation

▲ Material preparation and off line setup areas were separated and far from each other.

Inventory

▲ Too many partial license plates returned to stock.
▲ Too many setup carts with a single license plate on it in the *kanban* area.

Process Improvements

After a detailed analysis of the process, the following waste elimination goals were set:

▲ Install a computer application to automate the entire process, perform it on one station with one operator (instead of three), and cut lead time by 50 percent.
▲ Provide bins with more than three modules/separators to avoid marking materials for the 10th module.
▲ Move NXT offline setup area to the same area material is received from stockroom.
▲ Integrate LCR measurement into setup verification to prevent the license plate from being verified on the feeder if it does not pass electrical test. This eliminates problems with wrong component placements and redundant LCR checks.

▲ Provide white background on the desk surface where operators place components for the LCR measurement.

▲ Ensure that a unique license plate does not change for a particular piece of material to avoid repeated measurements of the same license plate and eliminate the need for placing a "pass" sticker on the license plate.

▲ Place another feeder bar code label on the opposite side of the feeder for easy and faster off line setup verification.

▲ To solve the issue with kitting enough license plates for part numbers of multiple feeders, identify 13 part numbers across 27 assemblies that are split. Based on historical work order run data, estimations of the number of license plates and quantity levels that need to be maintained in the small supermarket were made so that material could be placed alongside the line. Kitting is now a combination of central stockroom kitting plus kitting materials for these 13 parts at the supermarket.

▲ The supermarket is caged with a single entry point. The line lead operator will have control over the supermarket, picking parts and putting away parts.

The application performs the following component verification activities:

▲ Checks whether license plate is marked as defective due to recent customer feedback.

▲ Checks whether the license plate is lead-free if the job is for a lead free assembly.

▲ Determines which license plate is first to be set up on the machine.

▲ Performs component test: electrical (LCR) test if data is available, or if it isn't, a markings test if markings are available on the component, or manufacturing part number (MPN) verification if markings are not available either.

▲ Verifies that component belongs to the setup.

▲ Performs MPN data verification, i.e., checks whether feeder type matches required feeder type by MPN, whether feeder rotation matches MPN feeder rotation, and whether height matches MPN height. If MPN validation fails, which may affect machine program, black setup label is printed to warn process engineers that machine program change could be required.

DELIVER

Lean Case Study: Lean Supply Chain Reduces "Fat" by 13 Percent*

Customer's Challenge

Fujitsu Services is a leading European information technology services company, with an annual turnover of £2.46 billion ($3.59 billion) and over 19,000 employees in 20 countries. Its Sourcing & Supply Services operation provides purchasing and supply services for its major customers. In addition, its Technical Integration Centre (TIC) offers IT engineering, configuration and repair services, recycle and disposal solutions. These are both supported by a warehousing and distribution facility that provides secure bonding and storage.

With customers in areas such as banking, government and defense, Fujitsu's Supply Chain operation has to run without error or delay. However, in 2006 it faced some serious issues as Paul Fraser, Head of Logistics, Fujitsu, explains, "Quite simply we were failing. Delivery to customers on time was down to 95 from 99.9 percent, with significant costs being incurred for rearranged engineer visits and penalties for late installations.

"In addition, picking errors were running at over 4 percent, which meant having to rework many orders. Morale was low, with 14 percent absenteeism. Productivity was running at a minus figure, SLAs were being missed and we had to pay for off-site storage to handle 1,000 pallets of kit as our own 141,000 sq ft warehouse was full! In short, our reputation was badly damaged—customer expectations were low, complaints were increasing and new and key contracts were in serious risk of being lost. Something had to change significantly to improve the situation."

Fujitsu Solution

Drawing on the experiences of world class companies who use Lean operational practices together with Fujitsu's own unique Sense and Respond approach to continuous service improvement, Fujitsu's Supply Chain embarked upon a program of change utilizing elements of Lean, *kaizen*, and Six Sigma.

Source: www.Fujitsu.com

"We had to remedy what was going wrong before we could enhance the service we offered," says Paul Fraser. "So we analyzed the situation and found that we had a significant amount of customer-owned stock, much of it aging, and kit that needed to be recycled, but we had no instructions about what to do with it. In fact, it turned out that there were almost 90,000 units of redundant stock taking up space and just getting in the way physically and logistically, so it was inhibiting efficiency, morale was low and accidents were happening." Fujitsu realized that it needed to change the way it did business with its customers, so that it was clear about their needs, what had to be done to meet them and what was wasteful. Paul Fraser adds, "The most important thing is to talk to your customers and understand what is really of value to them, so that you can decide what to do away with without affecting customer service."

So, Fujitsu initiated a Lean program, called "The TIC Way," through which it developed a vision of where it wanted to get to and how it would measure its progress and success. This was captured in a detailed Transformation Plan, covering everything from leadership and processes, through to inventory controls and management. Fujitsu also created core team of employees to drive the change process. The team was built around the people directly involved in either identifying or creating the issues that need to be addressed, because of their key role in the problem solving sessions.

The rigid, command and control style of management previously used in Logistics has also been replaced. Today a more empowered workforce is instrumental in developing strategy and external relationships and involved in the problem-solving techniques used to understand and address the root cause. In particular, Fujitsu is using Visual Management techniques to monitor performance against targets, ensure actions from problem solving sessions are put into place and to monitor ongoing results once solutions are applied. These measures are openly displayed in each department used as a "Communication Hub," with teams being encouraged to contribute to the discussion and resolution and general updates. This information is then used across Sourcing & Supply Services to create awareness and initiate change across the wider capability.

"The introduction of Lean has had a positive effect on the way we work. Our workforce has become more involved with the processes and is taking more responsibility for the results we aim to achieve. Lean is also bringing the different departments together to work Leaner and more effectively," comments Tony Huddart, Operational Shift Manager, Fujitsu.

Paul Fraser agrees, "The vision we have is simple: get it right first time and with minimal touch points. Thanks to the use of Lean techniques our people now share that vision and the long term view of Supply Chain and Fujitsu Services and are passionate about the success and indeed the journey we have taken. Our next steps are to understand and address a further 30 to 40 percent non-essential time wastage through double or even triple handling and poor work flow, so that our customers get even more value from our services."

Benefits to Our Customer

Through the use of Lean operational controls Fujitsu has:

- ▲ Enhanced customer service—the identification and resolution of key operational issues enables faster and higher quality service delivery, which has seen customer satisfaction increase by 10 percent from 6.8 to 7.8 out of 10.
- ▲ Increased staff satisfaction—active involvement of staff in the change process has improved morale and absenteeism has fallen from 14 percent to just 3 percent, largely because there are fewer injuries.
- ▲ Improved resource usage—productivity has increased by over 23 percent and has enabled headcount to be reduced by over 14 percent, with the development of staff skills allowing them to work in other parts of the business.
- ▲ Reduced costs—has delivered 13.6 percent cost savings on a budget of £12.6 million, saving of £1.7 million in the first year. In particular, supplier deviations have been reduced by 18 percent, inventory is down by 19 percent, the need for off-site storage has been removed as redundant stock has dropped by over 90 percent to just 7,500 units, and costs of third-party handling and transport have been reduced.
- ▲ Created new revenue opportunities—better use of resources has enabled Fujitsu to increase volume throughput by 10 percent and develop new service offerings, such as engineering and workshop repair facilities.
- ▲ Enabled continuous improvement—involving and empowering people has created a sustainable internal capability, which is focused on continually identifying and permanently eradicating problems and waste.

"Through the use of Lean we have restructured our operation and processes to ensure the work flows and relationships with suppliers and customers are enhanced and we are more flexible within a highly controlled environment," comments Paul Fraser. "As a result, we have a very capable back-end to a very efficient and proficient front-end capability in the procurement of equipment and the delivery of engineering services, such as our break-fix and recycling capabilities.

We are now a leader in our field, because we understand what our customers want – and can deliver it."

Our Approach

First developed in the manufacturing industry, Lean is a systematic approach for identifying and eliminating waste or unnecessary activities through continuous improvement of the product or service in response to customer needs.

Paul Fraser says, "The involvement, inclusion and development of our people is key as it enables new processes to be put in place much more quickly than would otherwise be possible, because they have an understanding of the overall strategy. We've now got a team of people who question, recommend, advise, and are always looking for involvement in customers' needs. In fact, because we can now tap into the experience of our people in many cases we know what our customers want before they've recognized it themselves. And that mean they have pride in what they do, which is enabling the cultural change necessary for sustainable success."

Our Expertise

Fujitsu has been delivering consistently high levels of service to organizations across the United Kingdom for more than 30 years. Its business is helping its customers realize the value of information technology through the application of consulting, systems integration and managed service contracts. As such, its support infrastructure is constantly being developed and refined to stay ahead of the demands of new technologies and evolving business practice, so that services can be delivered effectively and economically.

DELIVER

Lean Case Study: Lean Logistics—Goodyear's Automated Warehouse Puts Customers on Top*

In manufacturing, the only thing as vital as the flow of product is the flow of information. Supply chain management is a critical component of doing business, and at The Goodyear Tire & Rubber Company, it's a priority. It's also one of the factors that made the company a global success story and North America's number one tire maker.

Paul Fledderjohann, Goodyear's Manager, Process Engineering, North America Tire (NAT) Supply Chain, is keenly aware of the distribution challenges. He is also well informed as to what tire manufacturing facilities are doing, not just to survive, but also to compete in an ever-evolving marketplace. It's a dilemma Goodyear faced in 2005 as it set its sights on improvements to the warehouse distribution system at the Goodyear facility in Fayetteville, North Carolina.

The Challenge

With a modernization project going on upstream from the warehouse, it was clear the Goodyear plant would require more than a retrofit for the current manual processes used for tire distribution. With the high number of SKUs, manual sorting capabilities had reached capacity, and Goodyear wanted to protect its workforce from the risk of injuries. It was also essential to have a Supply Chain Deployment strategy that offered real advantages to customers.

Since the expansion on the manufacturing side required half the existing warehouse space, Goodyear decided to build a new facility to handle tire storage and distribution. The challenge now was finding the right automation system, and the right provider to design and install it.

"It was a very strategic decision based on proposals received from the Bid Request and the supplier's past performance," said Fledderjohann. "We'd worked with RMT Robotics to build a fully-automated warehouse at the Goodyear plant in Lawton, Oklahoma, and we felt that solution would be a good system for Fayetteville."

---

*Source: RMT Robotics; www.rmtrobotics.com

In the last decade, Canada's RMT Robotics had established a solid track record with Goodyear, installing more than 30 robots at 6 of Goodyear's tire manufacturing facilities in North America. As the creator of large, high-velocity gantry robots and integrated systems, RMT understood the performance demands that Goodyear would require in Fayetteville.

The Concept

The challenge for Goodyear's NAT Supply Chain and RMT Robotics was to create a system for Fayetteville that could ship more tires directly to customers, and do it more accurately. As well, Goodyear wanted a system that could achieve payback very quickly.

The robotic distribution system that RMT proposed for the Goodyear plant featured 12 gantry robots working simultaneously to handle every aspect of warehouse distribution, from the time the product arrived from manufacturing, all the way through to sequential loading onto a trailer for customer delivery.

"The robotic gantry 'direct ship' solution is ideal in tire distribution," says Bill Torrens, Director Sales and Marketing at RMT Robotics. "In today's lean manufacturing and warehousing environment, it's a competitive advantage to have automation that can sort, temporarily stage, then ship tires directly to customers on demand. It not only reduces labor costs, it also keeps inventory levels low and customer response high."

RMT's "direct ship" system was capable of managing the entire system inventory and had no difficulties coping with the demands of a high-SKU environment.

"There could be up to 1,000 different SKUs in one month," said Fledderjohann. "As well, the Fayetteville facility can produce 55,000 replacement tires per day. To stock, stage and ship them from a deployment standpoint is very, very difficult and manually intensive. Our new system does it all automatically, first by identifying each tire, and then remembering that identification as it stacks, sorts and deploys to customer requirements."

Torrens describes the "direct ship" concept as palletizing avoidance. "Fed by final finish, the system maintains a large, dynamic picking inventory on the floor under the gantry, then ships it out in trailer loading sequence." He adds, "All of this without ever having seen a person, pallet or rack location. While the system is able to palletize to feed longer term storage, it's the high percentage that flows from final finish direct to trailer that makes the solution such a success."

In the fall of 2005, RMT Robotics was commissioned to begin the project. The work was completed in nine months, and after system optimization, Goodyear declared the installation phase completed in early May, 2006.

"It went even more smoothly than the Lawton installation," confirms Fledderjohann. "As a test, we processed about 30,000 tires per day for about a month and a half. When we finally went inside the gantry to do a physical count of the tires after processing 1.3 million tires, the discrepancy in inventory was basically nonexistent. That's very, very impressive."

The Future

Today, the robotic distribution system handles 100 percent of tire volume from the Fayetteville plant. The Goodyear Supply Chain Team confirms that, with a few modifications, the system is well suited to cope with any future production increases.

"I'm amazed at how well the technology is embraced," said Fledderjohann, referring to the pushback that sometimes occurs with the introduction of new technology. "This investment is good for the plant as well as for our customers."

Goodyear also recognizes the benefits to the supply chain that links customers with the Goodyear facility in Fayetteville. Now, with a fully automated system that also offers timely, accurate information on every tire regardless of where it is in the warehouse and deployment cycle, it's a real competitive advantage.

Goodyear's NAT Supply Chain is confident the new Automated Warehouse Facility will make it more attractive for customers to receive tires from Goodyear. This technology allows Goodyear to fully integrate production related systems to supply chain systems. That means better customer fill rates, better deployment of product, and ultimately the ability to offer more attractive solutions.

It's the kind of payback that Goodyear is counting on, and it's also a ringing endorsement for RMT Robotics. "It's been a good company to work with," said Fledderjohann. "They've delivered on what they promised."

"From the beginning, Goodyear and RMT Robotics have shared a common vision for the future of tire distribution," says Douglas Pickard, President, RMT Robotics. "It has benefited both companies, and the pioneering spirit of the relationship has resulted in solutions that fundamentally re-define the tire distribution model."

DELIVER

Lean Case Study: Lean Logistics—Want to Manufacture More Savings in Logistics?*

Just as fishermen learn how to go to where the fish are, savvy business executives go to where the ideas are.

That's just what happened at the nation's leading frozen seafood brand, Gorton's. In 1998 executives at this Gloucester, Mass., company read the book, *Lean Thinking*, by James Womack and Daniel Jones. With Vice President of Operations David Weber championing the cause, Gorton's began applying Lean concepts to its seafood processing practices. The goal: eliminate waste and wasteful practices while adding value to the process.

And although manufacturing would be the first area to learn about Lean, Gorton's plan was to adopt the practice everywhere.

"We adopted a culture of Lean throughout our entire business and that also meant spreading the message beyond our four walls," says Jeff Whiteacre, operations value stream manager. "Logistics was intentionally integrated from the very beginning. However, there was need to start internally and achieve success within Gorton's before we rolled out the plan to our suppliers and service providers."

With that, Gorton's developed and implemented training, and went to work. Soon, Whiteacre says, the results were dramatic. From 1998 to 2001 Gorton's reduced raw and finished material inventories by 50 percent, reduced its raw materials warehouse space and dramatically reduced operations time and handling by replacing individual bags of ingredients with truckloads of ingredients more directly fed into line operations.

By coordinating the efforts of logistics, purchasing, operations and sales, Whiteacre notes that one Gorton's plant was able to ship more than 90 percent of what it produced within 24 hours. Another plant increased its changeovers from 100 to 1,000 and reduced the amount of time to change the line from 1 to 2 hours to only minutes.

Source: By Bob Garrison. Reprinted with permission from *Refrigerated & Frozen Foods*, a BNP Media publication, copyright 2011. www.refrigeratedfrozenfood.com.

Results in hand, Gorton's was ready to spread the news. Moreover, it had the perfect forum for doing so. For years the company has hosted an annual operations conference for suppliers and service providers. The idea has been update these partners on Gorton's objectives and to foster a proactive dialogue.

Lean Thinking has been the conference theme for each of the past 3 years with Gorton's encouraging participation and offering support for training. This year's fall conference will go one step further with Gorton's honoring a supplier or service provider for their work with Lean (see sidebar, "And the winner is...").

Meanwhile, there is ample evidence that two of Gorton's logistics providers—AmeriCold Logistics and United States Cold Storage (USCS)—are applying Lean Thinking approaches to warehousing practices.

Like a Good Neighbor...

AmeriCold, Atlanta, operates a dedicated warehouse attached to Gorton's processing facility in Gloucester. After training with Gorton's personnel, Gloucester General Manager Gene Gallant led AmeriCold employees in a value stream mapping process to identify and scrutinize every warehouse activity. Next came a review of the Gloucester plant's practices and then, final discussions about where the two sides could jointly improve.

"We discovered—and since eliminated—a number of steps in our processes for handling product from the plant to the warehouse," says Kirk Hoover, AmeriCold vice president for lead logistics services. "When we looked at how we receive a pallet and acknowledge it, we found that there were duplicated steps and unnecessary administrative paperwork."

This year finds AmeriCold extending Lean practices to several of its distribution centers nationwide. Hoover notes, meanwhile, that DC improvements target more physical activities such as order picking and pallet handling on the dock and in the freezer.

Back in Atlanta, Hoover says AmeriCold has been applying Lean concepts internally and talking about the program with other processor-customers.

As far as AmeriCold's relationship with Gorton's is concerned, Hal Justice, executive vice president of lead logistics services, says Lean Thinking has its applications for public warehouses but fits even better in a dedicated warehouse strategy.

"To get the most from Lean the two facility managers—and the two companies—need an open and frank relationship. It requires a willingness to trust, share confidential information and talk about where costs are. It will become a win-win relationship only when the two facility managers realize that the object is to reduce total supply chain costs, not to reduce costs on one side by pushing it to the other."

Coast-to-Coast Coverage

USCS, Cherry Hill, N.J., has worked with Gorton's since the mid-1980s. USCS's Union City, Calif., warehouse first provided West Coast storage and distribution services for Gorton's. Over time, USCS's Fort Worth, Texas, site also has become a major partner in supplying the Southwest.

Notes Jerome Scherer, vice president of sales, marketing and government affairs, "Gorton's began speaking of Lean Thinking about four years ago at their annual conference. Then they started to promote Lean as an efficiency improvement tool for their suppliers and logistics partners. USCS began to employ Lean Thinking ideas in 2001."

Where better to try out a big idea than in a state known for big things. USCS Fort Worth took the lead role for value stream mapping to eliminate wasted effort and time in company practices.

Says Plant Manager Frank Monroe, "I'd recommend that any company dive into Lean. With representatives from every facet of operations involved you can better identify what's happening and map out strategies to eliminate waste. It's been an interesting and worthwhile effort. Any time you can drive out cost, it's good to do so."

Like AmeriCold, USCS has found ways to reduce administrative paperwork and increase productivity.

"We've moved away from paper by converting to radio frequency technologies," Monroe says. "That's eliminated redundant paperwork including, for example, pick slips and manual cases per man hour reports. Even the way we move around the facility has changed. A forklift does not move empty anymore. If it heads out to the dock to deliver a pallet, we make sure it returns with another pallet for storage.

"From my standpoint, we're just scratching the surface of what we can do and—while we're rolling out RF technologies to other USCS locations—we're also looking at more improvements everywhere from the dock to our offices."

Mike Goulart, Gorton's director of distribution, concurs.

"Each logistics partner brings its own creative approach to implement Lean techniques. There is no one right or wrong way to eliminate waste and provide value for the customer. Provided that our partners are looking at the entire value stream and seeing the whole picture, the results can be dramatic."

DELIVER

Lean Case Study: Warehousing Gets Lean*

While lean is long a fact in manufacturing, innovative managers at warehouses and DCs are now asking what lean can do for them.

Several years ago, OPW Fueling Components, a leading manufacturer of fueling products for gas stations and convenience stores, implemented lean manufacturing at its facilities in Cincinnati, Ohio.

With results like a 79 percent reduction in cycle time, OPW began to look at ways to apply lean concepts across the enterprise, including the finished goods warehouse.

"The lean concept has value, whether it's in the plant or the warehouse," explains Tom Ciepichal, vice president of operations. "We're going to identify and optimize those processes that add value for our customer. Then we'll reduce the processes that are non-value added or create unnecessary waste."

Just as in the factory, OPW is mapping out the processes associated with storing, picking and shipping finished goods, defining the steps that workers currently take to execute those processes, and analyzing how things are stored. They are also turning to technology, such as bar codes and supply chain management software (Glovia International), to provide better visibility into processes.

"To implement lean in the warehouse you need to look at how you can improve the visualization of the facility, set the stage for standard work, and look at the simple movement of goods," Ciepichal says. "For instance, we recently looked at how the layout of the warehouse impacts the smooth flow from picking and packing to shipping to optimize travel time."

Lean Warehousing?

Projects like the one at OPW beg the question: Is there such a thing as lean warehousing?

---

*Source: *Modern Materials Handling*, 12/1/2004, written by Bob Trebilcock, editor-at-large.

It's a good question. After nearly 50 years, lean manufacturing is a recognized discipline with well-defined best practices and many practitioners. Lean warehousing is nowhere near that stage.

"Lean warehousing is not yet a discipline," says Bruce Strahan, a partner with **The Progress Group**. "But people are asking what lessons from lean manufacturing we can apply to the warehouse."

For now, lean warehousing is a concept being embraced by manufacturers like OPW. These pioneers are exploring ways to translate their success on the shop floor into the other reaches of their businesses.

The state of lean warehousing might best be summed up by a recent logistics conference. There, an educational track was led by a plant manager who never once mentioned his warehouse. "That tells me that people are interested in the lean warehousing conference, but they haven't made much progress yet," says Jim Apple, another partner at The Progress Group.

There are some fundamental differences between the warehouse and factory that need to be taken into account. The most important of those, Apple believes, is variability and predictability. Even in a build-to-order environment, a plant manager knows what's going to run on the assembly lines for some period in advance. That allows for the synchronization of processes inside a facility, and ultimately across the supply chain in a lean operation. "As we've concentrated on shipping product on the same or next day, our ability to control demand and predict the mix of products that's going to be filled in the warehouse is low," says Apple. "Variability is what causes waste, excess labor and excess materials."

Still, Apple and others believe that while lean warehousing might still be in gestation, there are lessons to be learned from lean manufacturing that can be applied inside and outside the distribution center.

Inside the Four Walls

When it comes to warehousing, most talk about getting lean infers cutting back on personnel and inventory in order to do more with less.

But for lean manufacturing, as well as lean warehousing, that is only half the answer. The other half is getting agile.

"What is often missed is that leanness without agility is worthless," argues Stephen Parsley, principal engineer, **Daifuku America**. "To be lean simply means you've reduced the fat. To be agile means that your processes are flexible, scaleable, and above all, understood by all who are working with them so you can react to a change in plans."

Translation: lean is really about taking waste out of operations.

In fact, the original Toyota Production System identified "seven deadly wastes" that interfere with operations. The seven are overproduction, waiting, downtime, unnecessary product movement, excess inventory, unnecessary motion and defective products. The best practices that have come to define lean manufacturing, like faster setups and error-proofing, are designed to reduce the amount of non-value-added activity.

One of the initiatives at OPW illustrates that point. "In addition to the product we produce, we also purchase some finished product from other manufacturers," says Ciepichal. "We were storing, picking, and staging that product, combining it with our products, then shipping it to the customer. That was a lot of staging and sitting in queues."

Through better layout, design and storage, OPW created a plan to locate purchased product closer to the shipping dock. That reduces the amount of handling it takes to combine and ship it with products the company manufactures.

Another approach is to optimize those areas in the warehouse where there is predictability. "The variability in order mixes means we can't quite set up an assembly line for all of the warehouse," says Apple. "But we can identify those processes that have some stability."

Apple and Strahan recently worked with a leather goods manufacturer. One of the company's biggest problems is gift-wrapping during the holiday season, when 97 percent of their orders have just one item.

"What we realized is that even though each order might have a different item, we have a common category that will have a common processing requirement single-line orders that need to be gift-wrapped," says Apple. "That allowed us to develop an assembly line that will process those orders with half the labor of last season."

Warehouses can also apply the "touch-once" principle to their operations to reduce wasted motion. "The idea is that any time you touch a product just to move it, you're increasing your costs," says Ed Romaine, director of marketing for **Remstar International**. Creating buffer storage at the receiving dock to briefly hold inventory before it goes to a work area or is combined with other products to go out the door is one example of touch once, says Romaine. Another is to have vendors supply the product in finished shipping form with inner-packs, boxes and cartons that can be placed directly into the picking system. Picking to the shipping container to reduce the work at packing stations is another example.

Postponement is yet another best practice in the warehouse that enables Lean principles. "With postponement, you're not storing finished goods," says John Pulling, COO, **Provia Software**. "You're warehousing raw inventory in a work-in-process state, and then finishing it according to a customer's order. That especially makes sense if you have a lot of product variability."

Synchronization

Lean manufacturers are relying more and more on technology to synchronize the delivery of raw materials, parts, and components from suppliers with the product coming down the manufacturing line. That, in turn, is synchronized with the outbound transportation processes.

Manufacturers call this in-line sequencing, which means the parts that show up at the line are matched to specific product coming down the line. At the end of the process, they ideally go into the back of an outbound truck.

The emerging area of warehouse optimization software is one example of how in-line sequencing might translate to the warehouse.

"Lean is really about flow," says Pulling. "In a traditional warehouse, you pick orders and stage them on the dock until trailers show up. Optimization software allows you to schedule the arrival of trailers based on delivery dates, and then synchronize the release of orders to match the trailers arriving at the dock. By releasing orders in the proper sequence, you minimize intermediate storage and the labor associated with it."

There are other strategies to optimize the flow of product. Just as OPW reconfigured its warehouse to better deploy product, some companies are reconfiguring their supply chains to provide better customer service.

"We're working with one 3PL that is managing 350 stocking locations around the world to forward deploy spare parts for their clients," says Rob Sweeney, vice president of solutions engineering for **Yantra,** a provider of WMS and other supply chain execution solutions. "The other thing we're seeing is that end users are putting less emphasis on traditional WMS [warehouse management system] functionality like putaway and location control. They want functionality to manage flow-through centers, cross-docking, and merge-in-transit operations. They really want us to be able to synchronize their inbound and outbound activities."

At the end of the day, the most important Lean principle may have nothing to do with handling or configuration.

"The most important lean best practice is not a technique but an attitude," says Parsley of Daifuku America. "It's a total commitment to continuous improvement, with a never-wavering focus on the elimination of non-value-adding activities."

OPW Fueling learned that lesson during the implementation of lean manufacturing, and has carried it over to its lean warehousing initiative.

"'Continuous improvement is the key to our success with lean,' says Ciepchal. 'Once you complete a successful initiative, you'll see areas of opportunity to start another. And if you fall short of your expected results, you go back to analyze what went wrong, and do it again.'"

DELIVER

Lean Case Study: McKesson Moves Medicine*

The Challenge

As the largest pharmaceutical distributor in the United States, McKesson Corporation must keep its medicine on the move—our health depends on it. The company operates three business units, the largest of which is its pharmaceutical unit that distributes prescription drugs, health and beauty care products, and medical supplies throughout the United States from 27 stateside distribution centers.

More than 25,000 neighborhood drug stores, retail chains and healthcare facilities count on McKesson to help them operate their businesses with the highest possible quality, safety and efficiency. The company serves some of the country's largest drug store chains, including Walgreen's, Rite Aid and Longs, as well as more than 3,000 pharmaceutical manufacturers and medical-surgical supply developers who count on McKesson to bring their much-needed products to market.

The medical industry is expanding at a rapid pace. For McKesson this means an ever-growing number of products to distribute and a constant increase in demand. With a large number of SKUs to distribute and relatively small distribution facilities, McKesson understands the need to make the most effective use of their space.

In addition, like all pharmaceutical suppliers, each order McKesson fulfills is comprised of many individual units or "pieces." Heavy piece picking can mean extensive travel for order selectors as well as increased labor requirements and costs for the company. McKesson also guarantees next-day delivery for its orders, which adds additional pressure for accuracy and efficiency.

The Solution

For help in maintaining the ideal configuration of its facilities and ensuring peak efficiency and order accuracy, McKesson turned to Manhattan Associates' Slotting Optimization solution. This solution enables the company to determine the best location for each piece of inventory in their

---

*Source: *Supply Chain Digest,* January 17, 2008.

warehouse according to McKesson's unique business criteria. With optimal product placement that is based on current and historical demand, McKesson is able to increase workforce performance, shorten order fulfillment cycles and maximize customer satisfaction.

> Manhattan Associates' Slotting Optimization solution has helped our associates achieve incremental productivity gains that have contributed to McKesson's continued position as the pharmaceutical industry's benchmark for logistics and distribution efficiency.
>
> KEVIN PATTERSON
> Vice President, Distribution Operations Support,
> McKesson Corporation

The Implementation

Minimizing travel distance for and increasing the accuracy of McKesson's order selectors involved the creation and maintenance of logical picking zones within each McKesson facility. Like items are grouped together (prescription drugs, over-the-counter items and other health and beauty care items) and fast-moving items within each zone are consolidated into small areas strategically located at the most convenient locations, such as at the front of the aisles or adjacent to a dedicated conveyor. Within each picking zone, product is allocated between carton flow and static shelving according to its movement. This means that average historical velocities and sizes of all SKUs in the area are considered and product is arranged in the best way for it to be both picked and packed.

Because McKesson's product mix and demand naturally fluctuate throughout the year, the company's slot optimization requires ongoing maintenance to keep the pick zones balanced and ensure optimal productivity. McKesson has a dedicated resource focused on slotting at each distribution center and all facilities participate in monthly slotting meetings to ensure an exchange of best practices.

The Benefits

Manhattan Associates' Slotting Optimization solution has enabled McKesson to increase its picking efficiency by as much as 15 percent at some facilities. One facility measured a 14.8 percent improvement in the same 3-month period over the previous year after adopting a re-slotting program. All facilities using the solution have experienced productivity gains, while also reducing headcount. A key site reduced its Rx (prescription) and OTC (over-the-counter) order selectors from 19 to 17 after re-slotting those areas and another facility reduced OTC headcount from eight to six by properly allocating SKUs to separate conveyor lines based on their movement.

Increased efficiencies have also meant improved customer service at many locations. One location has succeeded at reducing defects per million opportunities from 750 to 200—meaning customers are benefiting from a dramatic increase in order accuracy. Most importantly, however, is the fact that the company's use of Manhattan Associates' advanced Slotting Optimization solution means McKesson's clients can provide vital medicines to their customers in record time.

The Future

For McKesson, whether it is an investment in process (a growing number of McKesson employees are Six Sigma certified) or technology, the investment's ability to create and measure efficiency is critical. The company's initial success with Manhattan Associates' Slotting Optimization solution certainly passes the test. Plans are currently underway to investigate integrating McKesson's labor management system with Slotting Optimization software for enhanced optimization results.

DELIVER

Lean Case Study: Norfolk Southern—Giving Customers More Value for Their Transportation Costs*

Last year's skyrocketing fuel prices made many of us look at transportation in a new light. As fuel prices rose, so did the prices of seemingly unrelated items, like groceries or building supplies. Often we don't think about transportation—at least not until any problems that affect the transportation industry start hitting us in our pocketbooks. But in truth efficient transportation is what makes our consumer economy work. Without it, goods couldn't be moved from where they are produced to where they are consumed. And as both production and consumption go global, the transportation industry plays an ever more crucial role in making sure the goods get where they need to be on time.

Norfolk Southern Corporation (NYSE: NSC) is one of the nation's premier transportation companies. Its Norfolk Southern Railway subsidiary operates approximately 21,000 route miles in 22 states and the District of Columbia and serves every major container port in the eastern United States, while also providing superior connections to western rail carriers.

The company employs 31,000 people and has 13 classification (hump) rail yards, 8 major system locomotive shops, 14 smaller division locomotive shops, and a total of 26 fixed facilities handling the servicing needs of these locomotives between runs. Norfolk Southern operates the most extensive intermodal network in the East and is North America's largest rail carrier of metals and automotive products. Railroads are also responsible for 75 percent of all automobile transport. The fastest growing segment of the business is in moving containers to terminals, where they are picked up by the customers.

People depend on Norfolk Southern's ability to get products where they are supposed to be on time, every time. When freight is shipped, the customer already has capital tied up in that freight, and that money can't

---

*Source: Mike Caldwell, TBM Senior Management Consultant, TBM Consulting Group, www.managingtimes.com, Q1-09.

be freed up until the customer has the product in hand and can put it on the shelf to sell. Late arrivals aren't just an inconvenience; they cost money.

Tracking Time and Money

In the freight business, time is the key. The time you've promised to deliver the customer's products to him. The time it takes to manage trains in rail yards. The time needed to service locomotives to keep them running at peak performance. The ultimate goal is to keep the trains running on time, which is very important from a customer point of view.

The greatest *kaizen* effort at Norfolk Southern has been geared toward locomotive performance and maintenance. This was straight *kaizen* work performed in the maintenance shops of the Mechanical Department in order to decrease the amount of time a locomotive is out of service. The railroad has about 3,800 locomotives. The lifespan of a locomotive is between 20 and 40 years. The majority is used for over-the-road freight hauling for their first 15 to 20 years and then are "cascaded down" to the rail yards to switch cars around the terminals. Similar to the airline model, federal regulations require that each locomotive receive increasing levels of maintenance inspections at regular intervals: daily, 92 days, 184 days, 368 days, 36 months, and 60 months. Understandably, the longer interval inspections require more time in the maintenance shop. Additionally, each locomotive will receive two to three complete overhauls in its lifetime. Locomotives are also inspected and fueled daily, although they don't need to go to maintenance shops for daily inspections, nor do they require fuel at every stop. Still daily inspections and fueling do take time, and that time must be accounted for.

Clearly, keeping those engines running must be a top priority for a business that relies on on-time delivery.

According to Gerhard Thelen, vice president of operations planning and support, additional business can be gained by keeping as many locomotives running as possible and is one way for the company to increase profitability. "We want to spend as little time on maintenance activities as possible, without sacrificing quality," he says. "Any time cycle time can be reduced, it helps the customer." Reducing the amount of time required to maintain the locomotives could also translate into a need for fewer locomotives to do the same amount of hauling that's being done now. Or the company can use the same number of locomotives and increase business. Either way, keeping the locomotives running means providing their customers with the service they have come to expect from Norfolk Southern.

> The vision: To be the safest, most customer focused and successful transportation company in the world.

According to Mark Smyre, manager of quality process improvement, the company has seen up to a 20 percent reduction in dwell time as a result of its *kaizen* activities (dwell time is the amount of time the locomotive spends in the maintenance cycle). Although you might think that this would mean that the Mechanical Department is releasing all of its locomotives back out onto the tracks sooner, in reality, some are going back out sooner, but for others, the company has chosen to dedicate the time savings gained to identify and fix items that can lead to potential failures.

"Unfortunately some locomotives don't stay out for the full 92-day cycle before needing unscheduled repairs," Smyre notes. "If we can take the time we've saved in our general maintenance cycle to perform proactive inspections, then we can identify potential defects and address them before they become an unscheduled maintenance problem, which ultimately saves ourselves and our customers time and money."

So how did they cut dwell time by 20 percent? They did it largely by focusing on the seven wastes—especially material transportation, wasted motion, waiting on parts, over processing (doing things that weren't necessary), and defects. Defects are defined as missing the identification of a potential failure and therefore not correcting it before the engine leaves the shop. "We don't want to compromise the size of our fleet as a result of missed opportunities," notes Smyre, "because we need the fleet to pull our freight." So it makes sense to address potential problems while the locomotive is already in the shop, and since time has been gained by applying Lean principles to the process, the maintenance crews can afford to spend extra time when necessary to ensure the quality of their inspections.

And that leads to the second part of cutting dwell time: standard work. "We also re-evaluated inspection procedures," Smyre says. "Standardization of our work processes was critical."

Rail Yards: They Aren't the Back Office

Being on time isn't all about keeping the locomotives running though. Another potential time sink is the rail yard. In rail yards, which operate in a hub-and-spoke manner, inbound trains are received; their engines are moved off for inspection or refueling, and the cars are shifted around to different trains with new engines and new destinations. As you can imagine, shifting rail cars and engines around and mixing and matching to create new trains, when everything must be confined to tracks, can be tricky and time consuming. In fact, according to Thelen, the greatest chance for delays to occur is when cars are being moved around yards. So it made sense to apply lean to the yards as well.

Generally when we think of business process *kaizen* (BPK) events, we think of events that help create smoother flow in the back office, but at Norfolk Southern, BPK is the method of choice for gaining greater efficiency in the yards.

Rail yards face a number of unique issues:

▲ Car inspection—finding, handling, and repairing cars in the yard
▲ Car handling—optimal utilization of crews for the most efficient movement of cars through the yard; extra crews must be justified
▲ Locomotive handling—efficiently moving engines to the maintenance and refueling areas
▲ Defect handling—what to do with damaged cars

"We are learning to view processes from the viewpoint of our customers, says Terry Evans, vice president of operations planning and budget. "To a customer, one-piece flow would mean that we would tow each railcar with its own locomotive directly to the customer's dock. We fully understand that we must run this business economically, but we must keep the customers view in mind. As we continue to identify process waste, we understand that there is an abundance of opportunities for improvement."

To date, Norfolk Southern has conducted 10 BPK events. The purpose of a rail yard BPK is to look at the entire flow of the process from cradle (inbound train) to grave (outbound train). They follow the same procedure as for any other BPK, mapping out the entire process using sticky notes on the walls. Doing so allows them to visualize all of the steps, including tasks, decisions, delays, and inventory. "We then identified the value-added steps in each process and tried to reduce or even eliminate some of the non-value-added steps," says Smyre. "As always, the value-added step is the one the customer is willing to pay for, and in the rail yard, the basic value-added service is getting the customer's car onto the right train and then getting that train out of the terminal on time."

Their objective for a BPK is to increase car velocity through the yards. By using the process map, it was easier to see where the potential bottlenecks were. "In one case," says Smyre, "we found that a train coming out of one particular yard could be better served by a different yard on the system, so we asked the Service Design Department to re-route that train out of a different terminal to better serve our customers."

A typical rail yard process might look like this: When a train enters a terminal, it is first parked in a receiving yard and the engine is cut off from the train and goes to the engine house for service. The body of the train is left parked in the receiving yard until its individual cars are "humped" onto "classification tracks." Once the various cars from the inbound train have been separated off and shifted onto the appropriate classification track, they are joined up with other cars bound for the same destination. At the opposite end of the classification tracks, the cars are pulled out and put into a "forwarding yard." From there they are dispersed onto outbound trains.

Rail yard BPKs have also allowed Norfolk Southern to take a close look at staffing alignment to make sure that they have people where they're really needed. This may seem to be common sense, but sometimes traditional staffing methods just remain in place and new people will be hired when in fact simply making sure that you have the right people in the right places is all that's really needed.

Another benefit that the company has gained from BPKs is the "viral spread of lean." "When we put together a team for an event, we include agreement and non-agreement people local to the terminal and also from outside the terminal who can come in and see things that the local employees may overlook due to familiarity," says Smyre. "Union leaders have been very supportive of our lean initiatives," he adds, "and from the point of view of our agreement employees, it instills ownership in our continuous improvement process."

"Having these events has not only gained us process improvements," continues Smyre, "it's also helped to familiarize our workers with their own terminals, some of which are large, complex places. And the outside workers invariably take the good ideas back and apply them at their own terminals."

Interconnectedness

While it may seem that Lean events at Norfolk Southern have been focused on two very different areas, they are in fact connected. Making the Mechanical Department more efficient has a direct impact on the Transportation Department, which is actually an internal customer of the Mechanical Department, because without healthy locomotives the Transportation Department can't get trains to their destinations on time. And if the Transportation Department can't operate efficiently, then the company might not be able to leverage the improvements in train maintenance to grow its business.

But ultimately Norfolk Southern has focused lean on its business for one main reason: efficient transportation means that customers get more value for their transportation costs. "Efficiency and on-time performance of the railroad has a direct financial impact on the customer that goes beyond the simple cost of transportation," notes Thelen. "And that also means profit for Norfolk Southern, because those companies with better service performance can charge more for that service. Again it's a matter of perceived value. Customers want their products on time and they want delivery times to be consistent; that is, if a product is coming from a particular location, they want it to arrive in the same timeframe every time. Our customers want to know that they will be receiving their shipments like clockwork. Norfolk Southern can guarantee that sort of delivery schedule thanks to its continuous improvement efforts."

Blue-Ribbon Standard

Norfolk Southern has already set the standard in rail transportation, and now it's doing the same with Lean and *kaizen*. "During a visit to Norfolk Southern, a Union Pacific mechanical officer who sat in on a portion of our *kaizen* activities marveled at the progress we have achieved in such a short period," says Smyre. And no one needs fear that the gains in efficiency have caused slippage in other areas: for 19 consecutives years Norfolk Southern has won the coveted Harriman Safety Award for having the safest employees in the railroad industry.

Lean has given the employees of Norfolk Southern a renewed pride in their work. Says Thelen, "I see the passion in our employees' eyes as they display their *kaizen* improvements, and those teams are displaying a sense of urgency. That will not be easily matched by the competition."

RETURN

Lean Case Study: Reverse Engineering—How to Gain Reverse Logistics Efficiency*

Here are three examples of how companies can rethink reverse logistics to gain greater supply chain efficiency and economy.

Forward thinking companies increasingly need to consider reverse. With so much attention, time, and capital spent on exploring ways to move the enterprise in new directions, what's left behind is often overlooked and under-controlled. Reverse logistics covers a wide array of services—from inspection, repair, and remanufacturing to consumer returns and after-market recycling. It can reduce waste and ancillary costs, drive sustainable best practices, or generate new revenue streams. It may include using inbound routing guides and core carrier partners to manage returns or outsourcing product lifecycle management to a 3PL.

CHALLENGE #1: Following a series of acquisitions, a retailer is managing reverse logistics regionally. Recognizing that a fragmented approach is creating redundancies, inefficiencies, and cost bleeds, it decides to adopt a centralized returns strategy.

SOLUTION: The company uses a demand-supply planning model to substantially reduce inventory by postponing unneeded repairs and focusing repair activity on meeting projected requirements for specific units. Additionally, integrating returns processing and repair operations reduces the return/repair cycle. Leveraging these efficiencies, the company increases control and centralizes returns processing; enables visibility to all inventory throughout the return/repair cycle; and purges unnecessary investment in buildings and systems to manage reverse logistics.

CHALLENGE #2: An e-commerce company expands by selling into brick-and-mortar retail outlets. As its logistics requirements grow, it struggles to efficiently manage warehouse space and labor. Managing fulfillment and returns poses an additional challenge.

---

*Source: Reprinted with permission from Inbound Logistics Magazine, November 2009. www.inboundlogistics.com/subscribe. Copyright Inbound Logistics 2009

SOLUTION: The company sells its warehouse and materials handling equipment, and outsources inbound and outbound distribution to a third-party logistics provider. It reduces its warehouse footprint and labor need by 50 percent, automating processes while improving space utilization. The company then reinvests the capital it recovers from selling the warehouse into growing its business.

CHALLENGE #3: A manufacturer dealing with sensitive, high-value medical parts is hampered by a lack of field inventory visibility. Inexact and non-automated processes for returning parts into its pipeline also create inventory management challenges. Delivering critical service parts to, and managing returns from, more than 1,000 field service technicians is rife with inefficiency.

SOLUTION: The company opts to work with a logistics service provider to manage a national network that includes 20 parts depots and a central distribution hub. Together with its 3PL, the manufacturer identifies key elements of the returns process that need improvement. As a result of these business process changes, the company increases visibility to partson-hand for field service technicians, dramatically reduces inventory costs, and centralizes all returns to a single location for better control. Reverse logistics becomes even more important when the bottom line drops, budgets cinch, and sales grow sluggish—when economy and customer service become paramount. Manufacturers are challenged to maintain high cost structures without risking lost sales due to poor customer service. Retailers, too, must focus on outward-looking forecasts to match marketing and sales efforts with demand. Overstock and returns are often unavoidable and they account for considerable expense. Some companies may rewire their internal infrastructure and work with logistics partners to manage the returns process; others completely outsource reverse logistics to reduce fixed costs.

APPENDIX B

Lean Opportunity Assessment

	Internal Communication	Score (1–5)
1.	Management communicates with all levels of the organization on topics regarding organization goals and objectives at least twice per year.	
2.	Employees are able to accurately describe the organization's goals and how their job contributes to the achievement of those goals.	
3.	Employees receive feedback through a formal process concerning problems found in downstream processes or from the customer.	
4.	Management encourages Supply Chain & Logistics employees to work in groups to address performance, quality, or safety issues.	
5.	Employees at the operations level understand and use common performance metrics to monitor and improve the production processes.	
6.	Problems in the Supply Chain & Logistics process are detected and investigated within 10 minutes of the first occurrence.	
7.	The concept of Value Stream Mapping is understood and all product famalies have been mapped and are physically segregated into the like process streams.	

Internal Communication Category Score = 0%

	Visual Systems and Workplace Organization	**Score (1–5)**
1.	The Distribution Center and Office areas are generally clear of unnecessary materials, items or scrap. Isles are clear of obstructions.	
2.	The Distribution Center floor has lines that distinguish work areas, paths and material handling isles.	
3.	All employees are aware of good housekeeping practices and operators consider daily cleanup and put away activities as part of their job.	
4.	There is a place for everything and is everything in its place. Every needed item, tool, material container, or part rack is labeled and easy to find.	
5.	Display boards containing job training, safety, operation measurables, production data, quality problems and countermeasure information are readily visible at each production line or process and are updated continuously.	
6.	Check sheets describing and tracking the top quality defects are posted and are up to date at each work station.	

	Visual Systems Category Score =	**0%**

	Operator Flexibility	**Score (1–5)**
1.	Operators are given formal training before doing a job on their own. Few defects or productivity related slowdowns are attributable to new or inexperienced operators.	
2.	Product/Component travel distances have been measured, analyzed and reduced by moving equipment and work stations closed together.	
3.	Equipment is "right sized" for the operation/process. They have the ability to change speed to match the takt time. No "monuments" are present in the process.	
4.	Operators are cross trained to perform other job functions and operators work in at least 2 different jobs each day.	
5.	Processes and equipment are arranged to facilitate continuous flow of work through the Distribution Center.	
6.	U-shaped cells have been designed and implemented to promote one piece flow where appropriate (e.g., light assembly area).	

	Operator Flexibility Category Score =	**0%**

	Continuous Improvement	Score (1–5)
1.	There is a designated champion and a clearly communicated strategy for continuous improvement in the facility with the necessary resources, organization and infrastructure in place to support the process.	
2.	There is a formal suggestion process in place to solicit ideas for improvements from all employees and to recognize their participation.	
3.	Employees have been trained in continuous improvement methods and have been affected by or participated in continuous improvement events.	
4.	Employees know the eight wastes, are actively involved in identifying wastes in their processes/areas, and are empowered to work to reduce and eliminate the waste.	
5.	Continuous improvement, *kaizen* projects/events are structured, planned and implemented. Successes are recognized and expanded throughout the facility.	
6.	Most improvements made throughout the Distribution Center and offices are made daily and involve little or no expense to implement.	
7.	Product/Process Value streams undergo examination for continuous improvement on a regularly scheduled basis.	

Continuous Improvement Category Score =   **0%**

	Mistake Proofing	**Score (1–5)**
1.	Employees have been trained in the basis of mistake proofing and there is a team responsible for analyzing production defects and identifying mistake proofing opportunities.	
2.	Mistake proofing devices and methods have been implemented or are being developed to eliminate the top production defects for each work area in the plant.	
3.	Parts, products and components have been analyzed to identify design opportunities to eliminate waste and improve productivity.	
4.	Operators are empowered to stop an activity when an error or defect is found or when they cannot complete their process according to the SOP.	
5.	Manual processes or tasks have been equipped with mechanical checks to aid human judgement whenever possible.	
6.	Equipment and processes are equipped with call (andon) lights or signals that bring attention to situations requiring assistance with a problem or the replenishment of supplies.	
	Mistake Proofing Category Score =	**0%**

	Quick Changeover/Setup Reduction	**Score (1–5)**
1.	Shift (Receiving, Shipping, etc.) and Wave startups are scheduled in advance and communicated to inform all workers that these events are on that day's schedule.	
2.	Shift and Wave teams are in place and have received training on setup time reduction procedures and are actively improving change over methods.	
3.	Shift and Wave startups are done frequently (Wave only) and typically take less than 10 minutes.	
4.	Startup/Wave time is visibly tracked and posted at each work station where work is performed.	
5.	Shift and Wave startup procedures are standardized and repeated in other areas of the plant. Standard procedures and checklists are visible and followed.	
6.	Special tools and equipment have been developed and implemented to reduce the time and labor involved in the startup process.	
	Quick Changeover Category Score =	**0%**

Quality - Inbound, Outbound, and Administrative	Score (1–5)
1. Zero defects from suppliers is a policy.	
2. The company quality system is effectively implemented and compliant with a national standard such as ISO-9000.	
3. FMEA (Failure Modes and Effects Analysis) is in place (Feedback, rootcause, etc.).	
4. Material Review Board/Discrepant material disposition is in place.	
5. Supplier quality systems are in place.	
6. Internal scrap loss is less than 1% of cost of goods sold.	
7. Returned material to vendors is less than 0.1% of sales.	
Quality I/B, O/B and Administrative Category Score =	**0%**

Supply Chain	Score (1–5)
1. Suppliers are involved in continuous improvement efforts with the company.	
2. Performance to delivery policy (on-time) is better than 98%.	
3. Quality performance of the suppliers exceeds 98%.	
4. Electronic communications with suppliers is used to trigger release of supplies under a *kanban* or VMI system.	
5. The company has regular input to the suppliers to improve design and performance characteristics of the supplied parts.	
6. Cost reduction goals with suppliers are documented and tracked.	
7. Sevice complaints with suppliers are resolved within 24 hours.	
Supply Chain Category Score =	**0%**

Balanced Flow	Score (1–5)
1.	There is an effort to level assembly, packaging, and administrative schedules by requiring suppliers to schedule frequent, smaller deliveries, over the period.	
2.	Changeovers in assembly, packaging and the office are made to support the concept of running to demand for all products, and not to support long production runs, WIP inventory buffers, or daily short ship emergencies, etc.	
3.	Takt time is known by all associates and determines the pace of assembly and packaging in the facility.	
4.	Assembly and packaging is facilitated through Value Stream Managers.	
5.	Processes on assembly and packaging lines or in cells (including the office) are balanced or leveled so the difference between cycle times of linked processes is negligible.	
6.	When demand volume changes, assembly, packaging, and administrative processes are re-balanced or redesigned to flex up or down the process cycle times to correspond to the new takt time.	
7.	When demand volume changes long term, supermarket and POUS levels are adjusted to meet the new takt time.	

Balanced Flow Category Score =   **0%**

Total Productive Maintenance	Score (1–5)
1. Maintenance team managers and workers have been trained in the basics of TPM.	
2. Machines and equipment have all necessary safety guards in place. Safety devices are in working order and equipment is locked out immediately when broken down or when otherwise appropriate.	
3. Preventive maintenance activity lists are posted in work areas and item completions are tracked over time.	
4. Accurate and visible maintenance records are kept up to date and posted nearby for all production and support equipment.	
5. Preventive maintenance activities are focused on increasing process utilization and minimizing cycle time variation.	
6. Preventive maintenance responsibilities are defined for both maintenance and production workers.	
7. Time is allowed in the daily production schedule for workers to perform their preventive maintenance and cleaning duties.	
Total Productive Maintenance Category Score =	**0%**

	Pull System (Assembly, Packaging, and Office)	Score (1–5)
1.	Each assembly and packaging cell, line, or process has displayed, visually, the target and actual hourly output, as well as the shifts production requirements and timing.	
2.	All assembly and packaging managers and supervisors have been trained in the principles and implementation of shop floor material pull systems.	
3.	Material flow or movement in the assembly and packaging area are based on the make one move one concept, or is dependent on individual pull signals, via *kanban*, etc. from downstream work stations as parts or materials are consumed.	
4.	Downstream processes are pulling material from upstream processes. Upstream production schedules are dependent on downstream use.	
5.	Packaging and assembly lines/cells are capable of adapting to changes in customer demand by changing only one production schedule at the pacemaker process.	
6.	Packaging and assembly supervisors are not motivated to produce more parts than the subsequent process require.	
	Pull System Category Score =	**0%**

	Standardized Work	Score (1–5)
1.	Standard operating procedures have been developed for each process or cell and are used to train operators.	
2.	Every Distribution Center and office process has its SOP posted within view of the worker performing the process.	
3.	The takt time (i.e., Demand Rate) for each product and/or service was used as the basis for the processing time for each operation and the process manning requirements.	
4.	The process of job design and standardization involves operators as well as support personnel.	
5.	Frequently repeated, non-value-adding operations in the facility such as setups, startups, quality checks, preventative maintenance, cleanup, etc. are visually standardized and updated.	
6.	Operators individually perform their processes according to the process sheets or SOPs and make few method or technique errors. Any errors are recorded and tracked.	
	Standardized Work Category Score =	**0%**

	Engineering	**Score (1–5)**
1.	Engineering personnel are aware, involved, and trained in Lean principles.	
2.	Systematic efforts are in place to reduce product variation and the number of items (part numbers) in the system.	
3.	Engineering has organized its activities along value streams.	
4.	Engineering processes are organized visually and the workplace shows evidence of visual indicators to show status of work.	
5.	Engineering processes have been balanced to create flow and reduce lead time within the engineering department.	
6.	Engineers routinely go to the location of a problem in production to assess the actual situation and communicate with the production operators to obtain their input.	
7.	Performance measures such as lead time and velocity are used to measure the department and establish goals for continuous improvement.	
	Engineering Category Score =	**0%**

	Performance Measurement	**Score (1–5)**
1.	Numerous and detailed financial reports have been replaced by a few key measures of enterprise performance.	
2.	Traditional cost accounting measures and individual/department efficiency measures have been replaced by value stream performance measures.	
3.	Performance results are communicated openly to all employees and are visually posted to show status and progress.	
4.	Employees understand how their individual efforts contribute to the overall results of the enterprise.	
5.	Individuals are rewarded for team-based performance rather than individual performance.	
	Performance Measurement Category Score =	**0%**

Customer Communication	Score (1–5)
1.	There is a standard system in place for collecting customer satisfaction information and data.	
2.	Customer requirements (including forecasts) are identified and communicated throughout the demand and supply chain.	
3.	Customer complaints are handled the same day they are received in under 2 hours and there is collaboration with customers to identify ways to reduce waste in the demand chain.	
4.	Customers have regular and systematic input into the design and functionality of the products they buy and there is an active Collaborative Planning Forecasting and Replenishment (CPFR) type function established for key customers.	

Customer Communication Category Score = **0%**

Lean Opportunity Summary and Graph

Category	Score
Internal Communication	0%
Visual Systems and Workplace Organization	0%
Operator Flexibility	0%
Continuous Improvement	0%
Mistake Proofing (Poka Yoke)	0%
SMED/Quick Changeover	0%
Quality	0%
Supply Chain	0%
Balanced Production (Assembly, Packaging & Office)	0%
Total Productive Maintenance	0%
Pull Systems (Assembly, Packaging & Office)	0%
Standard Work	0%
Engineering	0%
Performance Measurement	0%
Customer Communication	0%

Enterprise Characteristic
Consider the overall average of all categories
0%–20% Traditional Supply Chain & Logistics
20%–40% Getting started with Lean
40%–60% Lean Progress
60%–80% Value Stream/Lean focus/integrated Supply Chain
80%–90% Lean Continuous Improvement Culture
>90% World Class Lean Supply Chain & Logistics Management

Rating

Category Ratings

	0%	10%	20%	30%	40%	50%	60%	70%	80%	90%	100%
Internal Communication											
Visual Systems and Workplace Organization											
Operator Flexibility											
Continuous Improvement											
Mistake Proofing (Poka Yoke)											
SMED/Quick Changeover											
Quality											
Supply Chain											
Balanced Production (Assembly, Packaging & Office)											
Total Productive Maintenance											
Pull Systems (Assembly, Packaging & Office)											
Standard Work											
Engineering											
Performance Measurement											
Customer Communication											

REFERENCES

Chapter 1

Aberdeen Group, The Lean Supply Chain Report, http://www.aberdeen
.com (accessed 2011).

Heizer, J., and Render, B., *Operations Management*, 10th edition, Pearson,
Englewood Cliffs, NJ, 2010.

Schroeder, R., et al., *Operations Management*, 5th edition, McGraw-Hill,
New York, 2010.

SCOR Model—, http://www.supply-chain.org (accessed 2011).

Thomas, K., V.P. of Manufacturing at JDA Software, Inc., personal
interview, 2011.

Trunick, P. A., Continuing Education—Making the Right Selection, http://
www.inboundlogistics.com (accessed 2011).

Chapter 2

Bain & Company, http://www.bain.com (accessed 2011).

Chapter 4

Heizer, J., and Render, B., *Operations Management*, 10th edition, Pearson,
Englewood Cliffs, NJ, 2010.

McCreary, P., "Successful Lean Planning," *APICS Magazine*, May/June
2010, 38.

Servos, N., et al., "Implementing Six Sigma Principles in Reverse Logistics,"
proceedings of the 2009 Annual Meeting of Collegiate Marketing
Educators, February 28–29, 2009.

Worthen, B., Beware the Promises of Forecasting Systems, http://www.cio
.com (accessed 2003).

Chapter 5

Menlo Logistics, a Division of Con-Way, http://www.con-way.com (accessed 2011).

Chapter 7

Carleton, G., "Wringing Cost out of the Supply Chain," *World Trade Magazine*, February 2011.

Hayes, F. Walmart Takes Back its Supply Chain—IT in the Spotlight, http://www.storefrontbacktalk.com (accessed 2011).

Krizner, K. "Supply Chain Visibility and Efficiency Gets a Boost," *World Trade Magazine*, January 4, 2010, www.worldtradewt100.com, (accessed 2011).

Magretta, J., "The Power of Virtual Integration—An Interview with Dell Computer's Michael Dell," *Harvard Business Review*, Mar/Apr 1998, 81.

ManagementStudyGuide.com, Ecommerce and Internet Enabled Supply Chains, http://www.managementstudyguide.com (accessed 2011).

O'Reilly, J., "Supply Chain Velocity: Shifting into Overdrive," *Inbound Logistics Magazine*, http://www.inboundlogistics.com (accessed 2011).

Risen, C., "Dell takes on India," *World Trade Magazine*, September 2006.

Wisner, J. et al., Principles of Supply Chain Management—A Balanced Approach, *Cengage Learning*, 3rd edition, 2009.

Chapter 8

Bradley, P., "The Skinny on Lean," *Material Handling Magazine*, 2006 (accessed at http://www.dcvelocity.com, 2011).

Forger, G., "Menlo Gets Lean," *Modern Materials Handling Magazine*, November 1, 2005, 1–2.

Gaunt, K., "Are Your Warehouse Operations Lean?," *Universal Advisor*, (3): 1–2, 2006.

Menlo Logistics, Lean Logistics, http://www.con-way.com (accessed 2011).

Ryder Logistics, Five LEAN Guiding Principles, http://www.ryder.com (accessed 2011).

Chapter 9

Craig, T., International Lean Logistics—Beyond the Four Walls, http://www.ltdmgmt.com (accessed 2011).

Hexter, J., and Narayanan, A., S. McKinsey & Company and the U.S. Chamber of Commerce, *The Challenges in Chinese Procurement survey*, 2006, http:// www.mckinseyquarterly.com (accessed 2011).

Sardar, D., Cross Country Consulting interview, http://www.crosscountryconsulting.biz (accessed 2011).

Supply Chain Digest white paper, The 10 Keys to Global Logistics Excellence, 2006, http:// www.scdigest.com (accessed 2011).

Taylor, C., Five Reasons Why Global Logistics Is Moving from the Basement to the Boardroom, IBM Global Services white paper, 2006, 3–5, 14, http://www-995.ibm.com (accessed 2011).

Chapter 10

Dougherty, J., and Gray, C., *Sales and Operations Planning—Best Practices*, Trafford Publishing, 2006, http://www.grayresearch.com (accessed 2011).

Viswanthan, N., S&OP—Strategies for Managing Complexities with Global Supply Chains, http://www.aberdeen.com (accessed 2011).

Chapter 12

Aberdeen Group survey, The Transportation Benchmark—The New Spotlight on Transportation Management and How Best in Class Companies Are Responding, 2006, http://www.aberdeen.com (accessed 2011).

Aberdeen Group, The Lean Supply Chain Report—Lean Concepts Transcend Manufacturing through the Supply Chain, 2006, http://www.aberdeen.com (accessed 2011).

Aberdeen Group, Lean Manufacturing: Five Tips for Reducing Waste in the Supply Chain, 2009, http://www.aberdeen.com (accessed 2011).

Bjorklund, J., 10 Ways to Use ERP to Lean the Manufacturing Supply Chain, IFS software white paper, 2009, http://www.ifsworld.com (accessed 2011).

SAScom Magazine, "The Future of Forecasting Software," 2006, http://www.sas.com/news/sascom (accessed 2011).

Thomas, K., Senior Vice President Manufacturing at JDA Software Group, Inc. interview, 2011.

Chapter 13

AMR Research, Beyond CPFR: Collaboration Comes of Age, 2001, http://web.mit.edu (accessed 2011).

Fingar, P. et al., *Enterprise E-Commerce*, Meghan-Kiffer Press, 2000, http://www.firstmonday.org (accessed 2011).

Laudon, K., and Traver, C., *E-Commerce*, 3rd edition, Prentice-Hall, Englewood Cliffs, NJ, 2009, 77.

Sheffi, Y., The Value of CPFR, the MIT Center for Transportation and Logistics, 2002, http://web.mit.edu (accessed 2011).

Sheier, R. L., "Internet EDI Grows Up," *Computerworld*, 2003, http://www.computerworld.com (accessed 2011).

Supply Chain Management Review survey of CSC clients, http://www.SCMR.com, 2007 (accessed 2011).

Chapter 14

Cook, M., Philippe Hauguel and Roman Zeller Shaping Up Your Supply Chain, March 1, 2002, http://www.bain.com/publications/articles/shaping-up-your-supply-chain.aspx (accessed 2011).

Cook, M., Why Companies Flunk Supply Chain 101, Bain & Company survey and report, http://www.bain.com (accessed 2011).

Eckerson, W.W., Performance Dashboards: Measuring, Monitoring, and Managing Your Business, John Wiley and Sons, 2005, http://www.bpmpartners.com (accessed 2011).

Faldu, T. and Krishna, S., "Supply Chain Metrics that Measure Up—Building and Leveraging a Metrics Framework to Drive Supply Chain Performance," *Supply and Demand Executive Magazine*, May 2007, http://www.sdcexec.com (accessed 2011).

Srinivasan, M. M., "Seven Steps to Building a Lean Supply Chain," *Industry Week Magazine*, September 2007, http://www.industryweek.com (accessed 2011).

Chapter 15

Noe, R., *Employee Training and Development*, McGraw-Hill, New York, 2002, 26–28.

Chapter 16

Craig, T., President of LTD Management, interview, 2011.

Dogan, C., Gjendem, F., and Rodysill, J., "Fueling Supply Chain Transformation—Predictive Analytics Energizes Dynamic Networks," *APICS Magazine*, July/August, 2011, 41.

Fawcett, A. et al., Mastering Supply Chain Management, http://www.CSCMP.org, 2009 (accessed 2011).

Gatepoint Research and www.E2open.com, Supply Chain Benchmark Survey, 2009, http://www.e2open.com (accessed 2011).

Gordon, C., The Rise of the Supply Chain Officer, http://www.IMD.org, 2008 (accessed 2011).

Heizer, J., and Render, B., *Operations Management*, 10th edition, Pearson, Englewood Cliffs, NJ, 2010, 444–445.

Jordan, J., "The Data Analytics Boom," *Forbes Magazine*, 2010, last accessed at http://www.forbes.com (accessed 2011).

McCrea, B., "Taking a Global Approach to Education," *Supply Chain Management Review*, http://www.scmr.com (accessed 2011).

Schroeder, R. et al., *Operations Management*, 5th edition, McGraw-Hill, New York, 2011, 25–26.

Shacklett, M., "Supply Chain Software—The Big Spend," *World Trade Magazine*, 2010, http://www.worldtradewt100.com (accessed 2011).

Supply Chain Management Review survey of CSC clients, http://www.SCMR.com 2007 (accessed 2011).

Thomas, K., Senior Vice President Manufacturing at JDA Software Group, Inc., interview, 2011.

van Veen, J., "The Chain of Alignment," *APICS Magazine*, January/February, 2011, 35.

Viswanathan, N., and Sadlovska, V., Supply Chain Intelligence, Aberdeen Group, February 2010, http://www.aberdeen.com (accessed 2011).

Appendix A

A Lean-Six Sigma Duo for the Office, http://www.isixsigma.com; courtesy of Rath & Strong management consultants (accessed 2011).

Automotive Lean Supply Chain, http://www.georgia-associates.com (accessed 2011).

Caldwell, M., TBM Senior Management Consultant, "Norfolk Southern: Giving Customers More Value for Their Transportation Costs," http://www.managingtimes.com, Q1-09 (accessed 2011).

Cessna e-Procurement Project, http://www.esisinc.com (accessed 2011).

Garrison, B., "Lean Logistics—Want to Manufacture More Savings in Logistics?," *Refrigerated & Frozen Foods Magazine*, April 1 2003, http://www.refrigeratedfrozenfood.com (accessed 2011).

"Heinz Gets to One Number Forecasting," by SCDigest Editorial Staff *Supply Chain Digest*, October 20, 2008, http://www.scdigest.com (accessed 2011).

Lean Logistics—Goodyear's Automated Warehouse Puts Customers on Top, RMT Robotics, http://www.rmtrobotics.com (accessed 2011).

Lean Supply Chain Reduces "Fat" by 13%, http://www.Fujitsu.com (accessed 2011).

"McKesson Moves Medicine," *Supply Chain Digest*, January 17, 2008, http://www.scdigest.com (accessed 2011).

"Reverse Engineering—How to Gain Reverse Logistics Efficiency," *Inbound Logistics Magazine*, November 2009, http://www.inboundlogistics.com (accessed 2011).

"The Organized Office, http://www.obviousoffice.com (accessed 2011).

Trebilcock, B., "Warehousing gets Lean," *Modern Materials Handling*, 12/1/2004, (accessed 2011).

Vujosevic, R., Ramirez, J., A., Hausman-Cohen, L., and Venkataraman, S., Lean Kitting, http://www.optimalelectronics.com (accessed 2011).

INDEX

Content Group UK Ltd.
...on Keynes UK
UKHW020917250723
425739UK00002B/38

9 781265 629663